Assessing the Environmental Impact of Textiles and the Clothing Supply Chain

The Textile Institute Book Series

Incorporated by Royal Charter in 1925, The Textile Institute was established as the professional body for the textile industry to provide support to businesses, practitioners and academics involved with textiles and to provide routes to professional qualifications through which Institute Members can demonstrate their professional competence. The Institute's aim is to encourage learning, recognise achievement, reward excellence and disseminate information about the textiles, clothing and footwear industries and the associated science, design and technology; it has a global reach with individual and corporate members in over 80 countries.

The Textile Institute Book Series supersedes the former 'Woodhead Publishing Series in Textiles' and represents a collaboration between The Textile Institute and Elsevier aimed at ensuring that Institute Members and the textile industry continue to have access to high calibre titles on textile science and technology.

Books published in The Textile Institute Book Series are offered on the Elsevier web site at: store.elsevier.com and are available to Textile Institute Members at a substantial discount. Textile Institute books still in print are also available directly from the Institute's web site at: www.textileinstitute.org

To place an order, or if you are interested in writing a book for this series, please contact Matthew Deans, Senior Publisher: m.deans@elsevier.com

Recently Published and Upcoming Titles in The Textile Institute Book Series:

Handbook of Natural Fibres Volume 1: Types, Properties and Factors Affecting Breeding and Cultivation, 2nd Edition, Ryszard Kozlowski Maria Mackiewicz-Talarczyk, 978-0-12-818398-4
Handbook of Natural Fibres: Volume 2: Processing and Applications, 2nd Edition, Ryszard Kozlowski Maria Mackiewicz-Talarczyk, 978-0-12-818782-1
Advances in Textile Biotechnology, Artur Cavaco-Paulo, 978-0-08-102632-8
Woven Textiles: Principles, Technologies and Applications, 2nd Edition, Kim Gandhi, 978-0-08-102497-3
Auxetic Textiles, Hong Hu, 978-0-08-102211-5
Carbon Nanotube Fibres and Yarns: Production, Properties and Applications in Smart Textiles, Menghe Miao, 978-0-08-102722-6
Sustainable Technologies for Fashion and Textiles, Rajkishore Nayak, 978-0-08-102867-4
Structure and Mechanics of Textile Fibre Assemblies, Peter Schwartz, 978-0-08-102619-9
Silk: Materials, Processes, and Applications, Narendra Reddy, 978-0-12-818495-0
Anthropometry, Apparel Sizing and Design, 2nd Edition, Norsaadah Zakaria, 978-0-08-102604-5
Engineering Textiles: Integrating the Design and Manufacture of Textile Products, 2nd Edition, Yehia Elmogahzy, 978-0-08-102488-1
New Trends in Natural Dyes for Textiles, Padma Vankar Dhara Shukla, 978-0-08-102686-1
Smart Textile Coatings and Laminates, 2nd Edition, William C. Smith, 978-0-08-102428-7
Advanced Textiles for Wound Care, 2nd Edition, S. Rajendran, 978-0-08-102192-7
Manikins for Textile Evaluation, Rajkishore Nayak Rajiv Padhye, 978-0-08-100909-3
Automation in Garment Manufacturing, Rajkishore Nayak and Rajiv Padhye, 978-0-08-101211-6
Sustainable Fibres and Textiles, Subramanian Senthilkannan Muthu, 978-0-08-102041-8
Sustainability in Denim, Subramanian Senthilkannan Muthu, 978-0-08-102043-2
Circular Economy in Textiles and Apparel, Subramanian Senthilkannan Muthu, 978-0-08-102630-4
Nanofinishing of Textile Materials, Majid Montazer Tina Harifi, 978-0-08-101214-7
Nanotechnology in Textiles, Rajesh Mishra Jiri Militky, 978-0-08-102609-0
Inorganic and Composite Fibers, Boris Mahltig Yordan Kyosev, 978-0-08-102228-3
Smart Textiles for In Situ Monitoring of Composites, Vladan Koncar, 978-0-08-102308-2
Handbook of Properties of Textile and Technical Fibres, 2nd Edition, A. R. Bunsell, 978-0-08-101272-7
Silk, 2nd Edition, K. Murugesh Babu, 978-0-08-102540-6

The Textile Institute Book Series

Assessing the Environmental Impact of Textiles and the Clothing Supply Chain

Second Edition

Dr Subramanian Senthilkannan Muthu
Director & Head of Sustainability,
SgT & API, Cheung Sha Wan,
Kowloon, Hong Kong

The Textile Institute

ELSEVIER

WP
WOODHEAD
PUBLISHING
An imprint of Elsevier

Woodhead Publishing is an imprint of Elsevier
The Officers' Mess Business Centre, Royston Road, Duxford, CB22 4QH, United Kingdom
50 Hampshire Street, 5th Floor, Cambridge, MA 02139, United States
The Boulevard, Langford Lane, Kidlington, OX5 1GB, United Kingdom

Library of Congress Cataloging-in-Publication Data
A catalog record for this book is available from the Library of Congress

British Library Cataloguing-in-Publication Data
A catalogue record for this book is available from the British Library

ISBN: 978-0-12-819783-7

For information on all Woodhead Publishing publications visit our website at https://www.elsevier.com/books-and-journals

Publisher: Matthew Deans
Acquisitions Editor: Brian Guerin
Editorial Project Manager: Ana Claudia A. Garcia
Production Project Manager: Joy Christel Neumarin Honest Thangiah
Cover Designer: Victoria Pearson

Typeset by TNQ Technologies

Contents

Introduction to sustainability and the textile supply chain and its environmental impact

1

1.1 Introduction

Sustainability is being practised for many years in various industrial sectors including textiles and clothing. Concern on sustainability is increasing in a rapid pace in the textiles and clothing sector. There are umpteen number of definitions one can find from the literature on sustainability, still one of the most referred ones is from the Brundtland Report, which was published in 1987. According to this report, '*Sustainable development is the kind of development that meets the needs of the present without compromising the ability of future generations to meet their own needs*'[1]. Definitions or meanings for sustainability differ from both the people who define and the contexts and also importantly differ from various industrial sectors. The concept of sustainability revolves around three important dimensions or pillars, namely, environmental, social and economic. Sustainability or the sustainable development has to be a holistic approach that considers all these three pillars together. The ideology and the concept of sustainability has to begin with a broad scope for the industrial level and should finally be narrowed down to the product level considering all the phases of a product's life cycle. Today there are many definitions or concepts that lack this holistic approach, which makes the whole exercise of sustainability defeated.

When it comes to textiles, sustainability is being practised for a while, and as of today, it is practised in the industry as one of the essential business means. It is rare to see a company or a brand that does not practise sustainability in its business agenda or policy. The awareness of sustainability in the textile industry is certainly there and, of course, the reasons and motives of practising the same differs. No product can be made without any environmental brunt in this industrial era; however, what causes the difference is whether the brunt is necessary and this can be at a bare minimum level of possibility. A sustainable textile product is one that is made with the holistic consideration of environmental, economic and social aspects in the entire life cycle of a textile product.

Every product begins its life cycle at the raw material extraction stage, i.e. the cradle stage, and passes through various other stages, namely, manufacturing, distribution and use, before the cycle ends at the disposal (grave) stage. All the stages through which the product passes have an impact on the environment, as every industry has a dedicated supply chain for the manufacture of products and each part of the supply chain is responsible for a range of environmental impacts. Every individual consumes and disposes of a large number of products on a daily basis so the environmental impact increases with population growth if sufficient resources are available to support production.

Assessing the Environmental Impact of Textiles and the Clothing Supply Chain
https://doi.org/10.1016/B978-0-12-819783-7.00001-6

When this is reduced to the micro level for a single group of products such as textiles, the problem becomes more acute. Although other products are also responsible for damaging the environment, textiles are particularly significant because of their wide range of use.

Consumers use and dispose of many textile products at different times according to their purchasing power and needs. The consumption and disposal of textiles therefore rises as the population grows and becomes more affluent. This chapter deals with the basics of sustainability and reviews the entire supply chain for textiles and the clothing sector in terms of various processes from fibre to finished products and their environmental impacts. It also investigates the environmental impacts of different stages in the life cycle of textile products from the cradle to the grave.

1.2 Environmental sustainability

Environmental sustainability is one of the main pillars of sustainability and it includes the consideration of all the aspects pertaining to the environment when producing a product, such as the brunt on the environment in terms of its resources consumption and polluting the environment itself.

Environmental sustainability includes the consideration of reducing the consumption of all finite resources such as raw materials, energy, water and so on. This also includes the usage of renewable resources while consuming the above-mentioned elements. Major drivers under environmental sustainability are

- raw materials,
- energy consumption,
- water consumption,
- waste water discharge or water pollution,
- soil or land pollution,
- emissions to air,
- greenhouse gas (GHG) emissions or carbon footprint,
- hazardous waste management,
- toxic and hazardous chemicals management, etc.

Environmental considerations need to be enforced throughout the entire life cycle of a product from the raw material stage to manufacturing, distribution and mainly consumption stage, which includes the consumer use and disposal stages. The awareness of the environmental brunt is increasing, and these days, environmental sustainability is diversified into many spheres such as energy sustainability or footprints, water sustainability or footprints and chemical sustainability or footprints.

1.3 Social sustainability

All the three pillars of sustainability are interconnected, and the environmental sustainability and social sustainability are connected with each other. Social sustainability is quite in terms of scope and it focuses on people and their well-being. It enforces the confirmation of basic necessities of people and this includes umpteen numbers of issues such as fair labour practices, gender bias, sexual harassment, education, equal opportunities, community development, child labour, work-life balance, health and safety, protection, human rights and wellness. This is also being practised in mandatory level in all the industrial sectors and most of the companies today have a Corporate Social Responsibility division.

1.4 Economic sustainability

Needless to say, this is the chief pillar that decides the financial success of a business and it includes financial costs and benefits. However, the distinction comes here in terms of sustainability vis-à-vis regular business connotation is economic sustainability is not about profit at any cost. Economic sustainability is centred around and is very much connected to the other two pillars. The main concern on sustainability related to economic sustainability is the process of implementing or practising environmental and social sustainability should not affect the economy, meaning the product should not be too much expensive just for the sake of practising the other two pillars of sustainability. This includes issues such as long-term planning, cost savings, productivity, living cost, development and smart growth and so on.

1.5 The textile supply chain: an overview

Textile products encompass a wide spectrum of applications such as apparel, industrial textiles, geo-textiles, agro-textiles and hygienic textiles. They have varied life spans according to their durability and purpose. The textile and clothing supply chain is therefore particularly complex, as even within a single sector, for instance, clothing, there are many segments in the supply chain. In most cases the supply chain is both highly global and decentralized.

The initial element of the textile supply chain is fibre production. This is followed by yarn and fabric production and ends in the apparel manufacturing process, after which the finished product is ready to be sent to the customer. Diversified production lines exist for similar clothing, based on factors such as the type of material used and the end product required. There are also numerous production techniques for different fibre types, yarn spinning systems and fabric and garment technologies. As the supply chain is complex, it is difficult to map the processes and study the environmental impacts.

A generalized product life cycle model for a typical textile or garment is shown in Fig. 1.1. This diagram illustrates the eight stages of a product's life cycle and enumerates the production processes and sequences of the supply chain. Stages 1–5 constitute the finished fabric production sequences, beginning with raw material preparation.

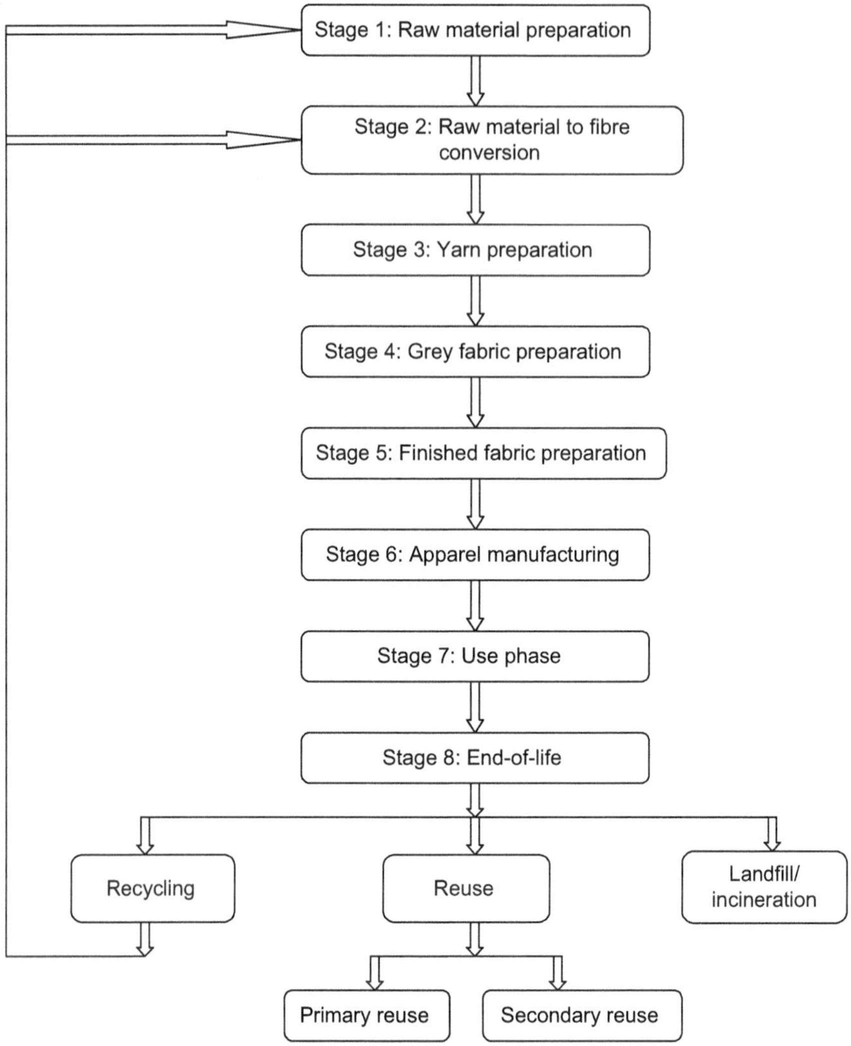

Figure 1.1 Generalized product life cycle model of textile products.

Stage 6 includes the garment manufacturing sequence, and stages 7–8 describe consumer use and various disposal scenarios.

Raw material preparation is the first step in the life cycle of textile products. There are two main sources of textile raw materials: natural fibre and man-made fibre. There are two sub-types of natural fibres: plant or vegetable (cellulose) and animal fibres. Typical examples of plant fibres include conventional and organic cottons, rayon, linen, hemp, jute, ramie and sisal. Wool, silk, mohair, cashmere, angora and alpaca are the main constituents of the animal fibre category. The production processes of

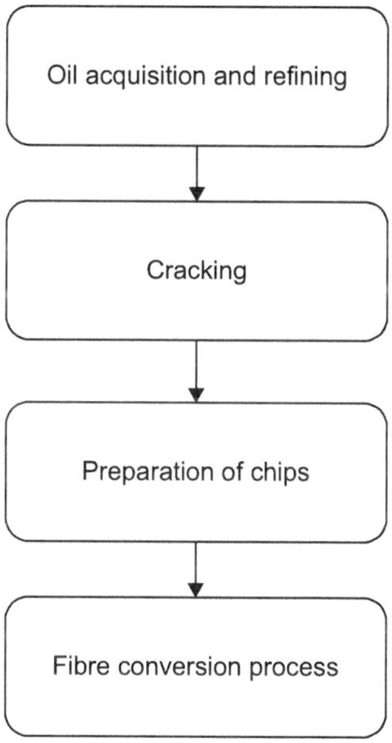

Figure 1.2 Synthetic fibre production.

natural fibres begin with cultivation, followed by growth and harvesting. After the fibre is obtained from its source, it will be transferred to a textile mill for further processes.

There are three types of man-made fibres: regenerated cellulosic, synthetic and inorganic. Regenerated cellulosic fibres are produced from the transformation of natural polymers and the fibres in this category include viscose rayon, acetate rayon, lyocell and modal. In man-made fibres of synthetic origin, the production process starts with the crude oil manufacturing process (the details are outlined in Fig. 1.2). There are many sub-processes between the crude oil manufacture and the preparation of chips and fibre manufacture. Only the most important processes are shown in Fig. 1.2. The principal fibres in this class are polyester, polyamides (Nylon 6 and 66), polyolefins and polyurethanes. Man-made fibres of inorganic origin include glass, carbon and ceramic fibres.

The second stage is the conversion of raw material to a spinnable fibre. Although the nature and number of processes will vary according to the fibre type, cotton is a typical example and is illustrated in Fig. 1.3. Examples of the cotton production processes are illustrated up to stage 5.

The third stage is the preparation of yarn from fibre and the fourth stage is grey fabric preparation. The processes involved in these stages are illustrated in Figs. 1.4 and 1.5. Stage 5 is preparation of the finished fabric, constituting several processes as shown in Fig. 1.6. Stage 6 is the garment preparation process from the finished fabric, shown in Fig. 1.7. Stages 7 and 8 include usage and disposal, which are mainly influenced by consumer behaviour coupled with the functional and ecological

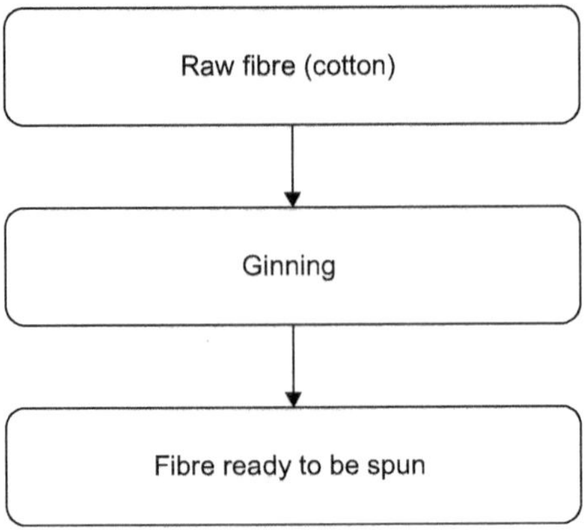

Figure 1.3 Raw material to spinnable fibre conversion process.

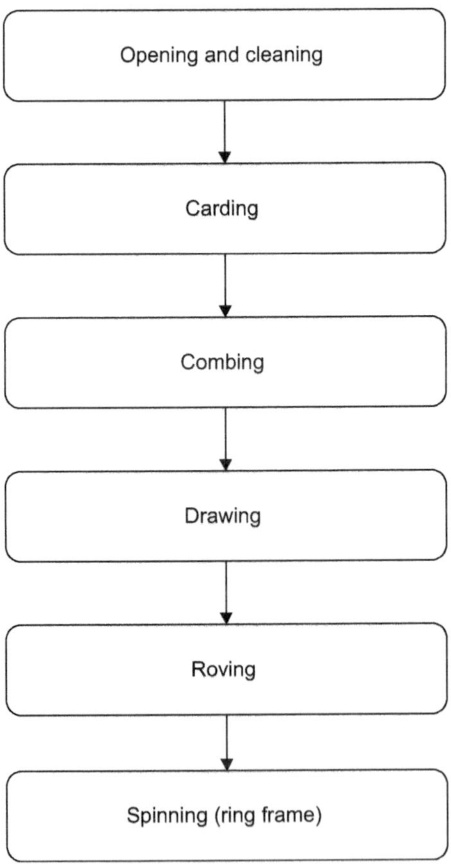

Figure 1.4 Yarn manufacturing process.

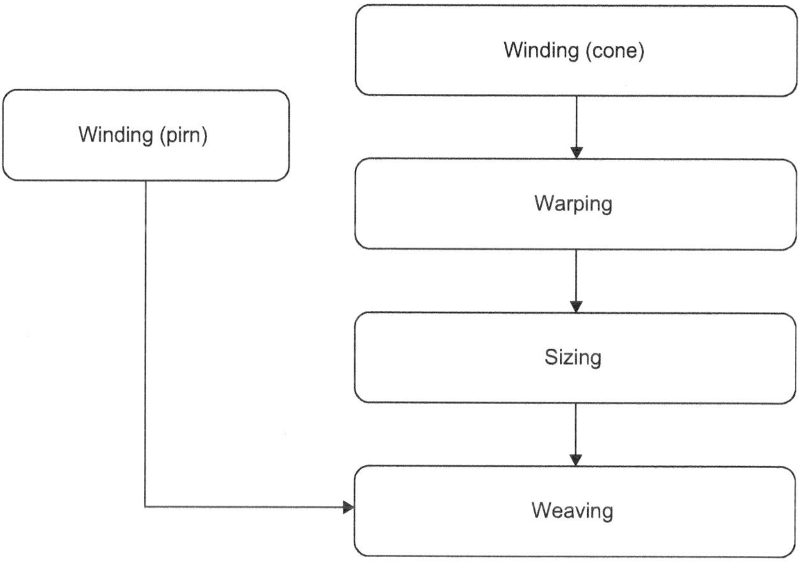

Figure 1.5 Grey fabric preparation process.

properties of the textile products. Details of these stages are given in Figs.1.8 and 1.9. The usage stage consists of the useful lifetime of the product and the maintenance required to sustain its shelf-life. The disposal stage includes the various possible destinations of a discarded product.

All the flowcharts in Figs. 1.1–1.6 describe the complete life cycle of textile products along with the various process elements embedded in each stage. There is no scarcity of literature explaining the processes involved in manufacturing a textile product from the fibre stage, and the key focus of this book is to enumerate the environmental impacts pertaining to the different life cycle stages. From the fibre stage to the disposal stage, there is a large input of resources, a high level of waste and emissions are produced and a large amount of energy is used in transportation. All these factors create local, regional and global environmental impacts.

Over its entire life cycle, a textile product requires the following inputs:

- direct usage of land to produce the fibres and the indirect use of land to build production facilities (even at the disposal stage, land is required for the option of disposing in a landfill and to build recycling/incineration facilities);
- freshwater from various sources for processing and cooling;
- energy from renewable and non-renewable sources for production and transportation;
- large amounts of pesticides, fertilizers, chemicals and other inventories;
- large amounts of packaging materials from different sources such as plastics and paper;
- inventories for the maintenance of machines.

This summary enables an overview of the textile and clothing supply chain and a brief introduction to its environmental impacts.

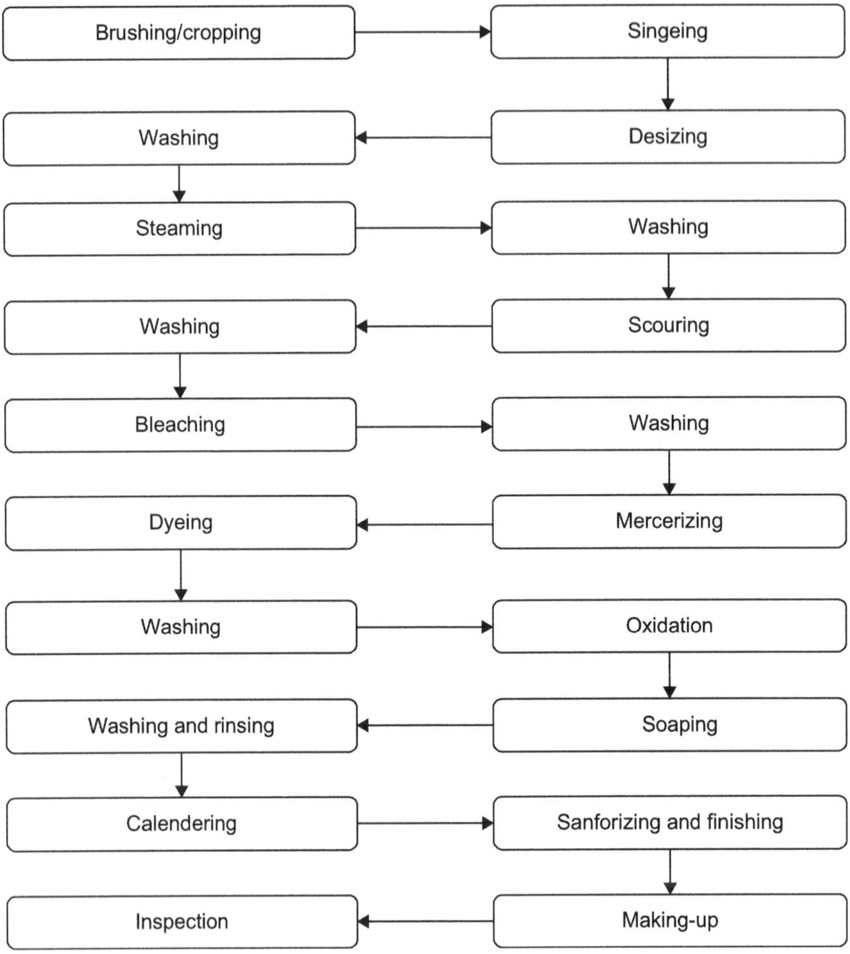

Figure 1.6 Finished fabric preparation process.

1.6 The production of natural fibres

In discussing the environmental impacts of textile products, it should be noted that confusion exists as to whether synthetic fibres or natural fibres are more environmentally friendly. The prevailing view is that natural fibres create lower environmental impacts, although it is not easy to come to a conclusion without assessing both synthetic and natural materials in light of the factors involved.

In 2011, Muthu and colleagues developed a unique scientific model for evaluating different textile fibres in terms of their environmental impact and ecological sustainability in order to calculate the environmental impact index (EI) and ecological sustainability index (ESI) of ten important textile fibres. In this model, the principal

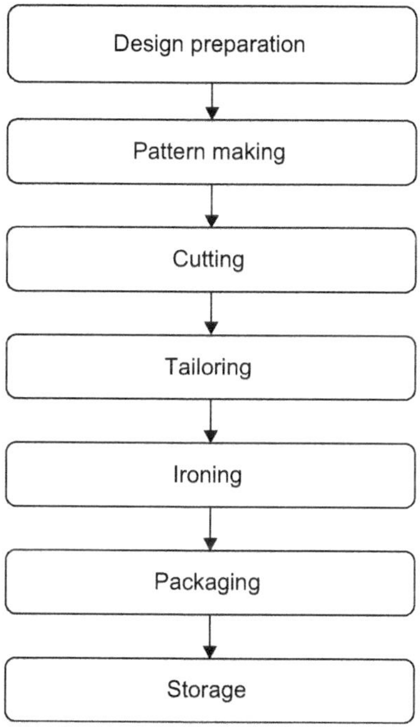

Figure 1.7 Apparel manufacturing production processes.

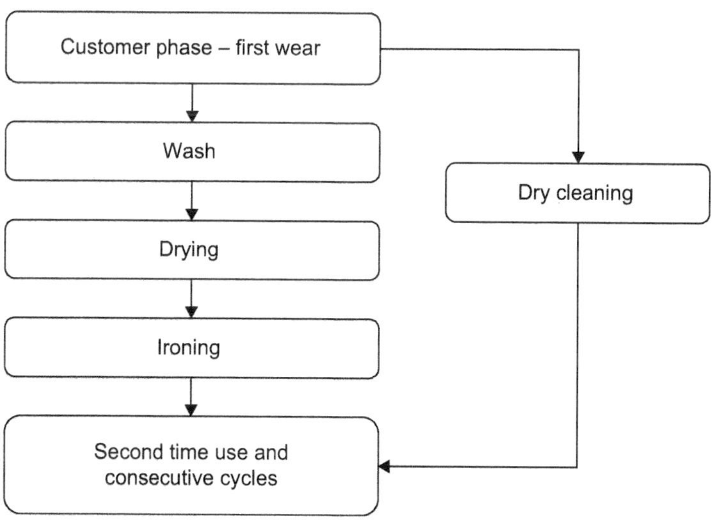

Figure 1.8 Use phase.

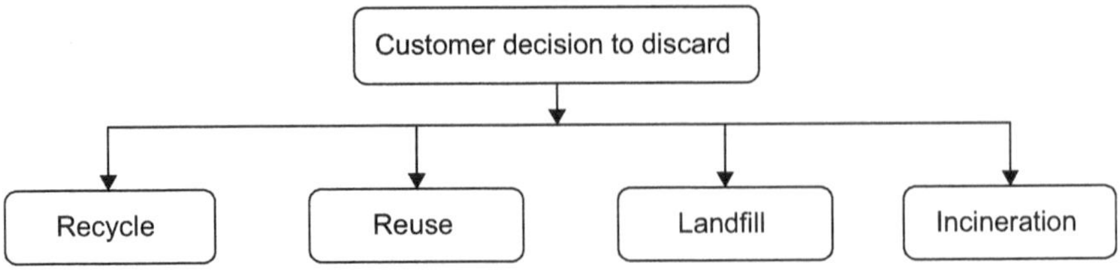

Figure 1.9 Disposal phase.

contributors to the environmental impact over an entire life cycle were taken into account:

- the amount of oxygen produced and carbon dioxide absorbed, offsetting global warming during the production phase of a fibre;
- utilization of renewable resources;
- land use;
- usage of fertilizers and pesticides;
- fibre recyclability;
- biodegradability;
- energy requirements;
- water requirements;
- GHG emissions.

Using this model, ten textile fibres were ranked in terms of their environmental impact and ecological sustainability. Organic cotton was found to have the smallest environmental impact with an EI of 11 and ESI of 71. Flax had an EI of 12 and ESI of 68. Conventional cotton and viscose had EIs of 16 and 19 and ESIs of 57 and 49, respectively. Polyester had an EI of 29.5 and ESI of 21. Acrylic was found to be the least preferred fibre in terms of environmental impact and ecological sustainability.[2]

Natural fibres may be categorized into two types: vegetable or plant fibres of cellulosic origin and animal fibres consisting primarily of protein. Animal fibres consist mainly of wool and silk. Wool fibres are further categorized into sub-types such as sheep, camel and goat. Vegetable fibres can be further grouped as[3]

- bast fibres such as jute, flax, ramie, hemp and kenaf;
- leaf fibres such as sisal;
- seed hairs such as cotton and kapok;
- leaf sheath such as banana and abaca;
- fruit fibres such as coir and pineapple.

Natural fibre production starts from agricultural production or animal source, followed by fibre extraction and processing for textile applications. The essential points for consideration in assessing environmental impacts include

- the amount of energy needed for production and the source of energy;
- the amount of pesticides/fertilizers used;
- the amount of water used and its source;

- the amount of other chemicals and consumables used;
- the amount, distance and type of transportation involved in the whole production process;
- the amount of packaging materials used;
- the type and quantities of pollutants released to air, water and soil;
- the amount of land used and the yield;
- the amount and density of waste produced.

Detailed production steps and their environmental impacts in specific important fibres are discussed in the following.

1.6.1 Cotton: conventional and organic

Cotton is the most widely used fibre in garment production. The cotton production chain requires the following key processes to provide fibre ready for spinning:

- sowing cotton seeds
- growth stage
- harvesting
- ginning
- baling.

This growth chain requires various inputs such as water, energy from both renewable and non-renewable sources, land, pesticides and fertilizers. All these are responsible for major environmental issues. The type and quantity of these resources differs between conventional and organic cotton. Many studies have pointed out that the production of conventional cotton is perceived as environmentally and socially hazardous, as it requires a higher usage of pesticides and fertilizers. According to the Environmental Justice Foundation (EJF) study, 2.5% of the world's cultivated land is used for cotton production and 16% of the world's insecticides are used on the crop. This is reported to be higher than the usage for any other single major crop. The World Health Organization (WHO) earmarked parathion, aldicarb, and methamidophos as the insecticides most hazardous to human health. These are among the top ten most widely used insecticides in cotton production. Aldicarb is reported to be so toxic that a single drop absorbed through the skin can kill a human being. An EJF study reported in 2007 that around 25 countries and the United States of America use aldicarb and that it has been found in the groundwater of 16 states.

Seven other insecticides used for cotton production are classified as moderately to highly hazardous by the WHO.[4] Information on Patagonia's website shows 10% of the entire production of agricultural chemicals are used for cotton production alone. It was also reported that conventional cotton produced in California consumes 6.9 million pounds of chemicals.[5] Among the 15 top pesticides, seven pesticides, acephate, dichloropropene, diuron, fluometuron, pendimethalin, tribufos and trifluralin, are used on cotton and are listed as 'possible', 'likely', 'probable' or 'known' human carcinogens by the Environmental Protection Agency of the United States.[6]

According to Laursen *et al.*, 1 pound of raw cotton in the United States consumes one-third pound of synthetic fertilizers. The excessive use of synthetic fertilizers is illustrated by the fact that it takes around 1 pound of cotton to make an average T-shirt.[7]

A survey on pesticides and fertilizers published by the National Agricultural Statistics Service in 2011 found that cotton used 92% of nitrogen, 67% of phosphate, 52% of potash and 42% of sulphur.[8] Nitrogen synthetic fertilizers are the most hazardous to the environment, as N_2O emissions have 300 times the effect of CO_2 in terms of GHG.[9–11] The United States Department of Agriculture (USDA) report showed that more than 2.03 billion pounds of synthetic fertilizers were applied to conventional cotton, making it the fourth most fertilized crop after corn, winter wheat and soybeans.[12] The potential risks of these pesticides and fertilizers were explained to farmers and their families.[4,11,13] All these environmental and human health hazards have resulted in the development of organic cotton, which does not use toxic chemicals and synthetic pesticides.

Some studies reported that conventional cotton requires more water than organic cotton,[14,15] whereas other studies reported the contrary.[16] Many studies reported lower energy consumption and carbon dioxide emissions for organic cotton than for conventional cotton. Energy use: Indian organic cotton—12 MJ/kg of fibre; US organic cotton—14 MJ/kg of fibre; and conventional cotton—55 MJ/kg of fibre. CO_2 emissions: Indian organic cotton—3.75 kg; US organic cotton—2.35 kg; and conventional cotton—5.89 kg CO_2 emissions per ton of spun fibre[17–19] were also reported in several studies.[14–19]

Organic cotton is superior to conventional cotton in terms of increasing biodiversity, mitigating climate change by elimination of intensive fertilizers, reducing water contamination and consumption, preserving soil quality and reducing energy requirements.[14–16]

A few LCA studies are worth mentioning here in terms of revised data available on the subject of conventional and organic cotton. In 2014, Life Cycle Assessment (LCA) of Organic Cotton Fiber was commissioned by Textile Exchange and PE INTERNATIONAL conducted the research. This study was based on primary data collected from producer groups located in the top five countries of organic cotton cultivation, namely, India, China, Turkey, Tanzania and the United States. The Life Cycle Assessment (LCA) model was set up using the GaBi 6.3 Software system, the functional unit being 1000 kg of lint cotton at the gin gate. According to the conclusions drawn, organically grown cotton has the following potential impact savings (per 1000 kg cotton fibre) over conventional cotton: 46% reduced global warming potential, 70% reduced acidification potential, 26% reduced eutrophication potential (soil erosion), 91% reduced blue water consumption and 62% reduced primary energy demand (non-renewable).[20]

Another LCA study[21] also included the Better Cotton Initiative cotton (BCI Cotton) in the LCA study of Organic and Conventional Cottons and this was chiefly focusing on the cotton cultivation practices of the three cottons, namely, better cotton, conventional cotton and organic cotton, in India. The information gathered from field observations and data collected from farmers were used to develop a model in the GaBi 8 Software released in 2017. The functional unit considered for the study was 1 metric ton of seed cotton at farm gate, for all the three systems, namely, better cotton, conventional cotton and organic cotton. The reference flow for all the three types of cotton cultivation systems was 1 metric ton of seed cotton. According to the study, the yield of each type of cotton per hectare played a major role. The LCIA results of the three cottons under this study are tabulated in the following[21]:

Impact category	Unit	Better cotton (yield: 1888 kg per hectare)	Conventional cotton (yield: 1938 kg per hectare)	Organic cotton (yield: 1755 kg per hectare)
Acidification	kg SO_2 eq.	12.41	12.68	0.57
Eutrophication	kg phosphate eq.	1.66	1.92	−0.02
Climate change	kg CO_2 eq.	688.00	680.20	338.50
Ozone depletion	kg R11 eq.	7.18E−09	6.90E−09	1.85E−09
Photochemical ozone creation	kg ethene eq.	0.17	0.15	0.05
Total primary energy demand	MJ	2.56E+04	2.55E+04	2.09E+04
Blue water consumption	kg	3.67E+05	3.44E+05	1.40E+05
Blue water consumption (including rain water)	kg	1.75E+06	1.71E+06	1.88E+06
Ecotoxicity	CTUe	1.17E+04	9.00E+03	1.41E−01
Human toxicity	CTUh	3.13E−07	1.82E−06	1.99E−10

CTUe, Comparative toxic units for ecosystems; *CTUh*, Comparative toxic units for humans; *MJ*, Megajoules.

1.6.2 Hemp and flax

Hemp and flax are considered to be the most significant sustainable fibres in the non-cotton natural fibre sector. Since the early 1980s, hemp has been used as a sustainable material for textile production and its advantages are as follows:[22–25]

- adapts easily to different climatic conditions,
- does not require pesticides and herbicides,
- has a modest requirement for fertilizer,
- does not require irrigation,
- suppresses weeds,
- disease-free,
- improves soil structure.

The production processes for hemp include ploughing, harrowing, fertilizing, sowing, growth and development, harvesting, cutting, seed extraction, retting, drying, scutching, baling, storage and fibre extraction.[23] Many life cycle assessment studies dealing with the environmental effects of hemp production have been reported.[25–29]

According to inventory details collected from sources, in addition to the work on global hemp production reported in a Life Cycle Assessment (LCA) study,[22] 1 hectare of hemp requires the following fertilizers: 85 kg of ammonium nitrate, 65 kg of triple superphosphate, 125 kg of potassium chloride and 360 kWh of electricity and other inputs for the scutching operation to produce 1000 kg of hemp fibres and other co-products.[22,25,30,31] The same study also reported 77.13 kg/ha nitrate loss to ground water and 2.55, 3.06, 0.31 and 1467 kg/ha emissions to ammonia, dinitrogen monoxide, nitrogen oxides and carbon dioxide, respectively, along with an assumption of 1 kg of methane released for each 150 kg of N applied as ammonium nitrate.[29,32–35] According to Carus and co-workers,[36] fertilizer input consumed 0.8 GJ/ton, farm machinery consumed 1.28 GJ/ton and the fibre processing stage and transportation consumed 0.94 and 0.8 GJ/ton of fibre, respectively.

Flax production processes consist of tillage, drilling, weed control, plant growth, dessication, harvesting, rippling, retting and decortication to produce flax fibres. This is followed by hackling, carding and spinning to produce flax yarns.[37–39] As reported in a study,[40] 972 kg of flax fibre yield was assumed per hectare. Nitrogen (N), phosphorus (P_2O_3) and potassium (K_2O) are the major categories of fertilizers used in the amounts of 40 kg/ha for N and 50 kg/ha each for P_2O_3 and K_2O (as per UK conditions).[37]

Different methods of tillage, i.e. conventional and conservational (having a lower number of passes over the field, so enhancing the surface residues to preserve soil and water loss), were compared in terms of energy use over 1 hectare of land. It was reported that 2.25 GJ of energy was consumed by the conventional method, whereas only 0.96 GJ was consumed by the conservational method.[41] The energy consumed in retting (bio and warm water) was reported as 0.48 and 0.03 MJ/kg, respectively. Scutching requires 0.53 MJ/kg and hacking, carding and spinning require 1.39, 3.94 and 22.9 MJ/kg of energy, respectively.[40] A small amount of herbicides, insecticides and fungicides were used in the cultivation of flax in amounts of 500 mL per 100 litres of water.[41]

1.6.3 Wool and silk

Farming and shearing are the main processes involved in the manufacture of raw wool fibre. Silk production begins with the cultivation of silk worms from eggs, and the main processes include silk worm rearing, cocoon production and the extraction of silk from cocoons (reeling process).

Environmental impacts pertaining to wool fibre production have been described in several studies.[18,42–44] The major impacts include

- soil compaction by the hooves of sheep;
- land clearing, which leads to loss of natural habitats, overgrazing on native pasture;

- methane (a GHG) generated by sheep;
- soil contamination due to the organophosphate used as sheep dip to control parasites, and organophosphate has also been found to be responsible for neurological problems in humans.

Research on energy consumption in merino wool production was carried out by Barber and Pellow in 2006.[43] They found that an average 890 MJ/ha and 230 MJ/stock unit were consumed on-farm; 50% of this was fuel and electricity, 36% was fertilizers, 9% was from capital and the remainder was contributed by agrichemicals. Barber and Pellow also presented an average inventory of resource inputs including direct and indirect energy needs. These details were collected from a survey of around 24 merino farms in a detailed LCA study of merino wool production. It was also reported that 66.6 kg of total farm production and 15 kg of wool were produced per hectare.[18]

The direct energy inputs were reported as 2.7 litres of diesel, 1 litre of petrol, 3.2 litres of diesel for contractors and 24.7 kWh of electricity per hectare. In the list of consumables (indirect energy inputs), 2.1 kg nitrogen, 4.4 kg phosphorus, 10.6 kg sulphur, 0.1 kg each of potassium and magnesium, 44.8 kg lime and 62.1 kg fertilizers were reported. This study also found the total farm emissions from the direct and indirect inputs listed earlier to be 59.6 kg CO_2.[18]

The major environmental concerns in silk production are the chemical fertilizers and pesticides used for the cultivation of mulberry trees, pollution generated by the wastewater released by the degumming process, increased water consumption in the silk fibre production process and land consumption for the plantation of mulberry trees.[45,46] (However, the amount of silk cultivated is lower than that of cotton.)

The positive impacts of sericulture are the reduction of salinity, the prevention of wind and water erosion and the improvement of air and water quality resulting from the planting of mulberry trees. However, the negative impacts must also be noted. Sericulture uses formalin and bleach powders as disinfectants and mulberry plantations require pesticides such as dyathin-M-45. The use of steam, fuels, water and chemicals in the reeling process also poses environmental threats.[47]

1.7 The production of synthetic fibres

More than half the world's garments are made from synthetic fibres. A study confirmed that in 2011, 79.1 million tons of textile fibres were produced, of which 61.3% were synthetic. Cotton accounted for around 31.2%; man-made cellulosic fibres, 6%; and wool, 1.5%.[48]

Synthetic fibres are criticized for their environmental impacts. The negative impacts associated with synthetic fibres[3,49] are as follows:

- obtained from non-renewable resources (depletion of fossil fuels is one of the major threats posed to the environment by synthetic fibres),
- require more energy in the production stage,
- produce considerably higher GHG emissions during the manufacturing phase,
- problems arise from the waste management,
- non-biodegradability,

- require a substantial amount of chemicals in the manufacturing process,
- emit toxic pollutants into different media across their life cycle,
- pose various health and toxicological threats.

General points to be considered in the environmental impact study of synthetic fibres are

- the amount of and source of energy needed in production,
- the amount of raw materials in terms of monomers used,
- the amount and source of water used in production,
- the amount of other chemicals and consumables used,
- the means of transportation and distance involved in the production process,
- the amount of packaging materials used,
- the variety and quantities of pollutants emitted to air, water and soil,
- the amount of waste produced and its management.

This section includes a review of the potential environmental impacts created by the four dominant synthetic fibres: polyester, nylon, polyolefins and acrylic. The manufacturing processes for synthetic fibres follow almost the same flowchart line, starting with the production of monomers and followed by polymerization, the spinning process, drawing, crimping, cutting and pressing into bales.

1.7.1 Polyester

Polyester is an important fibre in the synthetic category and is widely used in the textile and clothing sector. It is made out of purified terephthalic acid or dimethyl terephthalate and mono ethylene glycol.

The production of polyester fibre is an energy-intensive process (as high as 125 MJ/kg of fibre)[18] which produces high levels of GHG emissions. However, polyester has an advantage over natural fibres in terms of water consumption.[47]

Various studies have pointed out the different energy requirements of polyester fibre production, which lie between 104 and 127 MJ/kg of fibre. A study by Franklin Associates reported that 112 MJ of energy is required to manufacture 1 kg of polyester.[51] A study by Cherret et al. indicated that production in America has a higher energy consumption, i.e. 127 MJ/kg, than in Europe (104 MJ/kg).[28]

One of the important studies earmarked for the life cycle assessment of polyester is the LCA of a polyester blouse, which was carried out by the American Fiber Manufacturers Association in 1995. Although now outdated, this is an important study, hence its inclusion here. This study defined the functional unit as 1,000,000 occasions of wearing a polyester blouse. The polyester resin and fibre manufacturing process accounted for 9% of the total energy consumption, including packaging in the fibre stage, where the energy of the material resource alone consumed 3.6% of the total energy. The total industrial waste in the fibre manufacturing process, including packaging, accounted for 809.4 pounds.[52]

Kalliala and Nousiainen's[16] LCA study listed details of the production of polyester fibres. It showed 1 kg of polyester fibres to consume around 97.4 MJ of energy and 17.2 kg of water, with 2.31 kg of CO_2, 19.4 g of NO_x, 18.2 g of carbon monoxide,

39.5 g of CH emissions to air and 3.2 g of water emissions (chemical oxygen demand [COD]).

A study of the life cycle of a polyester garment reported that the resin needed to manufacture a polyester jacket weighing 500 g required 26.82 MJ of energy and emitted 1.55 kg CO_2, 4.14 g of NO_x, 2.38 g of particulate matter and 4.48 g of SO_2.[53]

According to the eco-profiles of a plastic industry project, 1 kg of bottle-grade polyethylene terephthalate (PET) requires 69.4 MJ energy and 60 kg of water and produces 2 kg of CO_2 emissions and 0.57 kg of solid waste. Water emissions are lower, for instance, 0.0014 kg COD was emitted.[54]

The polyester production process emits volatile organic compounds (VOCs), acetaldehyde and dioxins, which pose a severe threat to human health and the ozone layer. In addition, catalysts such as antimony that are used in PET production are carcinogenic.[55] Wastewater emitted from polyester processes consists of volatile monomers. Other by-products and solvents used in polyester production require strict control.[49]

1.7.2 Nylon

Nylon, which is known as polyamide, is another important textile fibre and was primarily intended to replace silk. There are many variants, of which the most common are Nylon 6 and Nylon 66. Nylon 6 is produced from caprolactam and Nylon 66 is produced from hexamethylenediamine and adipic acid.

The production process has a high level of energy consumption, which may reach 250 MJ/kg of fibre.[18] Even higher energy requirements, i.e. 262 MJ/kg, are reported by some references including the GaBi database (one of the commercial LCA software tools).[18] The production of nylon also creates nitrous oxide, which is a significant GHG.[42] At the end of its life cycle, Nylon 66 is very difficult to recycle and if burned, emits poisonous gases such as dioxins, nitrous oxide and hydrogen cyanide.[56,57]

According to the eco-profiles of a plastic industry project, 1 kg of Nylon 66 requires 138.62 MJ and 663 kg of water, whereas Nylon 6 demands 120.47 MJ and 185 kg of water.[58,59] Plastics Europe has listed a complete inventory of emissions for both Nylon 6 and 66. Important and significant parameters in that list include 6.5 kg CO_2 per kg of Nylon 66 and 5.5 kg CO_2 per kg of Nylon 6. Nylon 66 produces 18 g of sulphur emissions (SO_x as SO_2), 14 g of nitrous emissions (NO_x as NO_2) and 49 g of methane. Nylon 66 is also responsible for 15 g of COD, 3.6 g of biochemical oxygen demand (BOD) and 3.9 g of total organic carbon (TOC).[58] Nylon 6 emits 17 g of sulphur (SO_x as SO_2), 19 g of nitrous material (NO_x as NO_2) and 47 g of methane. Nylon 6 is also responsible for 3.6 g of COD, 0.56 g of BOD and 0.053 g of TOC.[59]

1.7.3 Polyolefins

Polyethylene and polypropylene (PP) are the most common fibres in this category of synthetic polymers. They are produced by the polymerization of ethylene and propylene, respectively.

There are several variants of polyethylene. Among them, low-density polyethylene (LDPE) and high-density polyethylene (HDPE) are common. LDPE has a density less than 940 kg/m. Polyethylene with a density of more than 940 kg/m is regarded as being HDPE.[60,61] About 1 kg of LDPE and HDPE production consumes 47 and 32 kg of water and 78.08 and 76.71 MJ of energy, respectively.[19,20] LDPE produces 1.7 kg of CO_2, 5 g of sulphur emissions (SO_x as SO_2), 3.8 g of nitrous emissions (NO_x as NO_2) and 16 g of methane.[60] HDPE produces 1.6 kg of CO_2, 4.1 g of sulphur emissions (SO_x as SO_2), 3.2 g of nitrous emissions (NO_x as NO_2) and 14 g of methane.[61]

PP consumes 73.37 MJ of energy and 43 kg of water for 1 kg of production and emits 1.7 kg of CO_2, 3.8 g of sulphur (SO_x as SO_2), 3.3 g of nitrous (NO_x as NO_2), 6.1.g of carbon monoxide and 12 g of methane.[62] A study by Barber and Pellow reported that production of 1 kg of PP requires 115 MJ of energy.[17] PP is not biodegradable or easily recyclable.[63]

1.7.4 Acrylic

Acrylic fibres are produced from polyacrylonitrile. Due to its warmth and wool-like feel, it is generally chosen as a cheap alternative for cashmere wool and is used in a wide range of textile applications such as socks, sweaters, gloves and home furnishings.

Laursen *et al.*[7] reported that the production of 1 kg of acrylic fibre requires 157 MJ of energy, whereas Barber and Pellow[18] reported 175 MJ/kg and the GaBi software database reported 194 MJ/kg of fibre; 1 kg of acrylic production requires 210 litres of water and emits approximately 5 kg of CO_2.[7]

The manufacturing process also utilizes toxic substances that require extremely careful handling. The storage, disposal and polymerization process for acrylic fibres emit toxic fumes that pose a threat to human health. Current regulations require acrylic production to take place in a closed environment where the safe disposal of fumes may be assured.[64,65] Acrylic is neither biodegradable nor easily recyclable.[63]

1.7.5 Viscose rayon

Viscose rayon is a semi-synthetic, regenerated cellulosic fibre and is important in the textile sector. Its production sequence is lengthy and consists of the following main processes:[66,67]

- dissolution of cellulose in caustic soda
- pressing
- shredding
- xanthation
- dissolving
- ripening
- filtering
- degassing
- extrusion-wet spinning
- drawing
- washing and cutting.

The production of viscose is energy intensive and emits higher amounts of GHG when compared with cotton production. Energy demands of viscose are as high as 100 MJ/kg[18,68]; 1 kg of viscose production requires 640 litres of water.[7] The process uses carbon disulphide as a solvent, which is highly toxic to human health and to the environment; 50% of unutilized CS_2 is released into the atmosphere, adding to the dangers described earlier.[68,69]

1.8 Spinning

The conversion of fibres into yarn is performed by the spinning process. The initial process for cotton is ginning, which is followed by opening and cleaning, carding, draw frame, simplex (roving formation) and yarn spinning. Modern spinning systems differ from conventional ring spinning and eliminate some of these processes. Different fibres require specific spinning systems; for instance, wool spinning utilizes processes and machines that differ from those used in cotton spinning. The initial process in the spinning of wool is scouring, which is followed by drying, dusting, mixing and oiling, carding, gilling, combing, pin drafting, roving and spinning.

Whatever system and techniques are used, the following details must be considered for their environmental impact:[70–73]

- transportation from farm to ginning facility;
- transportation from ginning facility to spinning factory;
- types and distances of internal transportation between different spinning departments;
- energy use and sources in the various processes;
- list of consumables used, such as lubricants, packaging materials and their disposal;
- amount of fibre waste created and its disposal;
- chemicals used and their disposal;
- amount of dust, short fibres and noise created;
- requirement for humidification systems;
- inventory of production accessories such as cone inserts, plastic ring cops, roving bobbins, card and draw frame cans and their disposal.

Dust is the major pollutant in cotton spinning and poses the risk of chronic bronchitis. Long-term inhalation of cotton dust may cause the respiratory disease byssinosis, and many studies report health issues pertaining to cotton dust from spinning.[73–80]

The chemical treatments prior to spinning, such as scouring and carbonizing, use substances including sodium carbonate, sulphuric acid, detergents, soaps and alkali. These are responsible for irritation of the eyes, nose and skin, and their effluents (creating BOD and COD) and solid wastes cause various environmental problems.[73]

The energy consumption of spinning systems used for cotton and other fibres differs and depends on many factors, including geographical and technological. A typical spinning plant having both ring and open end spinning systems is described in the following to illustrate the power consumption of various spinning processes:[81,82]

- blowroom—11%
- carding—12%

- draw frame—5%
- combing—1%
- roving—7%
- ring spinning—37%
- open end spinning—20%
- winding—7%.

In a typical spinning mill, 78% of energy is consumed by machines, 3% by lighting, 3% by compressors and 16% by humidification plants.[81,82]

The amount of energy consumed by spinning knitted and woven yarns is different. Woven yarns require more twisting and the production speed is therefore slower than that for knitted yarns. Finer yarns consume more energy and combed yarn requires more energy than carded yarn because of the additional processes involved.[82] Koç and Kaplan[83] calculated the energy consumption of combed and carded yarns of different counts used for both knitting and weaving purposes. As examples, carded yarn with a count of 37 Tex requires 1.34 kW h/kg for knitting and 1.62 kW h/kg for weaving. Combed yarn with the same count uses 1.38 kW h/kg for knitting and 1.63 kW h/kg for weaving.

The dust, particulates, solid wastes and noise generated during the cotton spinning process create a major environmental impact. Volatiles, acid fumes and effluents such as high solids, BOD and COD created in wool spinning also pose a threat to the environment.

1.9 Fabric manufacture

The fabric manufacturing process has many variants such as knitting, weaving and nonwovens. Knitting is a relatively simple process that uses a single machine to convert the yarn to grey fabric (except in the case of warp knitting, where warping is an additional process). Weaving consists of various preparatory processes, namely, winding, warping and sizing. Nonwoven manufacturing is an entirely different concept of fabric production directly from the fibre stage. The various processes involved in different fabric manufacturing technologies and the machines employed means that the degree of environmental impact created differs. The factors include

- transportation from the spinning facility to the fabric factory;
- the means of internal transportation and the distance between different departments of a fabric factory;
- the energy consumed in various processes and its source;
- list of consumables used, such as lubricants, packaging materials and their disposal;
- amount of solid waste created and its disposal;
- chemicals used and their disposal;
- amount of noise created;
- requirement for humidification systems;
- list of production accessories such as cone inserts, empty beam frames and their disposal.

Noise and solid waste are major environmental concerns in the knitting industry.[73] There is a scarcity of data on life cycle assessment studies for the knitting sector and for the hot spots within the process. A limited number of studies have dealt with energy consumption and carbon emissions in the production of knitted T-shirts. A study reported in 2009 showed the production of a T-shirt weighing 0.25 kg required 2.56 MJ of energy and released 0.16 kg of CO_2, 0.46 g of particulate matter, 0.96 g of NO_x and 0.99 g of SO_2 emissions.[53]

Environmental concerns related to weaving are greater than those arising from knitting owing to the increase in the number of processes and machines involved. A variety of weaving machines are currently available, each with its unique processing methodology, energy needs and environmental impacts. Irrespective of the type of weaving machine or technology used, the following issues require close attention:[73]

- dealing with the solid wastes of yarn and fabric scraps;
- solid wastes of size residues (applied during the sizing process);
- particulates and VOCs from the sizing process and particulates from the weaving and warping processes;
- water-based effluents from the sizing process;
- high-decibel noise from looms.

In the weaving process, size is applied to lubricate the warp and is later removed from the fabric in the desizing process. The agents used in this process are made of synthetic polymers or polysaccharides and result in a heavier load on wastewater in terms of increased COD. This accounts for around 30%–70% of the COD load.[84]

As with knitting, there is a dearth of life cycle assessment studies in the weaving sector. No studies have been reported on hot spots in the weaving process that might lead to a reduction of its environmental impact. Diversified results have been presented regarding energy consumption and it is difficult to derive the energy needs of weaving per unit weight of fabric. The Environmental Resources Management's study on Poly Lactic Acid (PLA) products shows the total weaving process to consume 12.60 kW h per kg of product as extracted energy (winding—1.19 kW h/kg; warping—3.80 kW h/kg; weaving—7.6 kW h/kg).[85]

A study by Kalliala and Talvenmaa[86] reported that approximately 5.4 MJ/kg of energy is required for weaving. A study by Turunen and van der Werf[40] pointed out that a range of 15–57 MJ/kg cotton is required in the process of weaving. Blackburn and Payne's towel study[87] towel study stated that the total weaving process consumes 10.6 MJ/kg of energy, of which winding accounts for 1 MJ/kg, beaming and sizing account for 3.2 MJ/kg and rapier weaving accounts for 6.4 MJ/kg of energy. The same value of 10.6 MJ/kg of fabric was reported by another study.[88]

A life cycle assessment of incontinence products published by Abena[92] reported that the production of 1000 kg of spun bonded nonwoven requires 611 cubic metres of water and 111.7 MJ of energy. It also found that the process of producing 1000 kg of nonwoven emits a long list of effluents to air and water, including 2.9 kg of CO_2, 8.2 kg of methane, 17.1 kg of SO_2 and 14.1 kg of NO_x. However, environmental impacts or life cycle assessment studies in the nonwovens sector are still in their early stages and few studies are available for discussion.[89–91] Nonwoven manufacturing processes consist mainly of web formation and condensation followed by a bonding process. VOCs, particulates and solid waste consisting of fibre waste and fabric scraps present the main environmental concerns.[73]

1.10 Finishing processes

The final process in fabric manufacture includes many divisions or sub-processes. There is no fixed finishing route for any fibre or any type of fabric and finishing processes will vary depending on the end use. The main processes include singeing, desizing, scouring, souring, bleaching, mercerizing, dyeing, printing, sanforizing, calendering and the application of other special finishes.

The finishing process creates air and water emissions and solid wastes, but it is the water emissions that are of primary importance and concern. This is mainly due to the large quantity of chemicals used throughout the process. The following points are important when considering the environmental impacts of the finishing process:

- production of various chemicals and other auxiliaries;
- maintenance and safe disposal of dyes, chemicals and auxiliaries;
- quantity of chemicals and dyes used per unit weight of textile;
- transportation type and distances from weaving to finishing factory;
- internal transportation types and distances between different departments of the finishing factory;
- amount of water used for processing and its source;
- amount of effluent discharged to different media;
- treatment of effluents in a wastewater treatment plant;
- safe disposal of sludge produced;
- energy used in various processes and its source;
- other consumables used in various finishing processes such as lubricants, packaging materials and their disposal;
- amount of solid waste created and its disposal;
- production of steam from boiler and its associated impact (e.g. procurement, storage and burning of firewood);
- production of hot air from appropriate sources and their associated impacts;
- quantity of hot air and steam used;
- production accessories for finishing processes and their disposal issues.

In addition to water emissions, energy consumption has a significant environmental impact. Textile finishing processes should be a subject of study for specific process-based LCA. On average, almost 1 kg of chemicals and auxiliaries are used per kg of finished textile. The highest environmental load arises from salts, followed by detergents and organic acids.[40] The quantity of effluent produced by a textile mill depends on various factors including the type of fibre and fabric material being finished. As with the combined effluent values from cotton and synthetics published by WRAP (Waste and Resources Action Programme), UK, woven fabric finishing produces 550–650 mg/litre of BOD and 850–1200 mg/litre of COD. Knitted fabric finishing produces 250–350 mg/litre of BOD and 850–1000 mg/litre of COD.[93]

The energy consumed by finishing processes varies according to the factors discussed earlier. Various studies have quoted different values. Blackburn and Payne's[87] towel study stated that 0.35 kg of dyeing chemicals are required per kg of fabric and that total dyeing and finishing consumes 30.8 MJ/kg. The breakdown of different operations associated with the total energy figure quoted earlier is as follows: singeing—0.2 MJ; brushing—1 MJ; desizing and washing—5.1 MJ; scouring—6.2 MJ; bleaching—2.8 MJ; mercerizing—5 MJ; dyeing by winch machine—5.7 MJ; padding to apply finish—3 MJ; and stentering to dry the fabric—1.8 MJ.[4]

Turunen and van der Werf's[40] study reported that the finishing process consumes 29–72 MJ/kg of finished textile.

1.11 Apparel manufacture

The final process in the clothing production chain is the manufacturing process, which is also called the garmenting process. This consists of various operations: spreading or laying, cutting according to the pattern, sewing and attaching interlining components to garments by heat pressing, ironing and packaging. The points to be considered in the environmental impact assessment are

- transportation type and distance from finishing factory to garmenting factory;
- internal transportation type and distance between different departments of the finishing factory;
- energy used in the garmenting operations;
- procurement of different accessories—distance and means of transport;
- dealing with scraps of fabric from various operations, especially waste from cutting;
- percentage of garments rejected;
- production of steam from boiler and associated impact (such as procurement, storage and burning of firewood);
- other consumables such as lubricants, packaging materials, transportation and disposal issues.

Studies dealing with the life cycle assessment or carbon footprint of textile products cover the processes of garment production. However, this has a relatively low impact when compared with other manufacturing processes of clothing production. A study concerned only with the life cycle assessment of clothing processes has been reported and found that the clothing is the cleanest of the manufacturing processes.[94] According to this study, clothing processes consume 2.472 MJ of energy, of which 49.8% is used in sewing, followed by cutting (29.6%) and packaging (20.6%).

1.12 Distribution and retail

This phase involves transport from the factory to the customer via the retailing process. The major impacts of this stage arise from the means of transportation and the distance involved. When compared with the other life cycle phases, the overall environmental impacts of transportation are small and result mainly from the energy profile. However, efforts should be made to minimize these impacts by reducing or avoiding long-distance transportation, particularly air freight, which contributes around 90% of the impacts arising from the distribution phase.[95] This has been emphasized in several studies.[49,95,96]

1.13 Usage and disposal

These are the final stages of the life cycle of a product. A life cycle assessment study is only complete and meaningful if it includes these two stages and is known as cradle to grave stage assessment. Life cycle assessment confined to the manufacturing stage is termed cradle to gate assessment.

These stages are largely controlled by consumers, whose attitudes play a significant part in deciding environmental impacts. The crux of a life cycle assessment lies in evaluating the impact of a product over its entire lifetime, of which the use phase is a significant part.

In the case of textiles, the use phase makes the largest contribution to the total environmental impact over the lifetime of a material. This will increase as the life span of the product increases. Products intended to have a longer life span, such as denims, will have greater use phase impacts than overall impacts. However, some products with longer life spans, such as jackets, will have less impact, as they require less frequent washing and drying.

The significant factors in the use phase are

- type of care needed for maintenance of the textiles,
- washing and drying methods,
- amount of water and chemicals used in washing,
- temperature of washing and drying,
- energy consumed in washing and drying,
- necessity of ironing and the energy consumed,
- frequency of washing.

It is difficult to draw generalized conclusions as to the impacts of the above-mentioned factors, as they will vary significantly according to consumers' individual preferences and habits and the country in which they live. Almost all the cradle to grave textile studies performed to date have emphasized the importance of reducing the use phase impact. Irrespective of the type of textile, this phase is generally responsible for up to 80% of the carbon footprint.[97] The consumption of energy, water, chemicals and other resources will vary depending on factors such as the type of product, fibre and end use.

To illustrate the differences in the above-mentioned points for various materials, a UK study on three entirely different products made from different fibres is discussed here.[96] This study considered a cotton T-shirt, a woven blouse made out of viscose and a nylon tufted carpet. The life cycle of the T-shirt required 60% of the total energy, i.e. 65 MJ for 25 washes at 60 °C, tumble drying and ironing. The viscose blouse, undergoing 25 washes at 40 °C and hang drying without ironing consumed 14%, i.e. 7 MJ of the total energy. The nylon carpet consumed 17% of the total energy in its use phase, i.e. 65 MJ over 10 years, including vacuum cleaning maintenance.

A study carried out in Germany considered a polyester jacket and cotton T-shirt. It reported that a T-shirt is washed 50 times over its lifetime, whereas a jacket will be washed six times during its life cycle. To meet the common functional unit of 100 days of wear, the T-shirt consumed 62.16 MJ in washing and 114.23 MJ in drying, whereas the jacket consumed 14.92 MJ in washing and 27.41 MJ in drying. Differing air emissions were also reported.[53]

Analyses of differing scenarios were included in most of the life cycle assessment studies and recommendations, such as changing consumer behaviour, reducing the number of washes, washing at a lower temperature and using natural drying, were made to reduce the use phase impacts on the total life cycle impacts.[53,87,95,96]

After the use phase the product reaches its end of life and may be directed to one of the following options:

- reuse for primary and/or secondary purposes,
- recycling,
- incineration with or without energy recovery,
- disposal to landfill.

As in the use phase, disposal is primarily decided by consumer behaviour. Each above-mentioned option has its own environmental impact or benefit. The first option is entirely beneficial, particularly if the product is reused by the original user. If the product is reused by another, the impacts arising from transportation, collection, sorting and reselling must be included in calculating the net environmental benefit.

The second option is recycling to create a new product for the same purpose or for secondary applications. Recycling involves breaking down the waste completely and using it to make a new product by mechanical, chemical or thermal means. The process of recycling generally involves the following four operations:

1. collection of waste,
2. sorting,
3. pre-treatment,
4. recycling (closed loop, open loop, incineration, landfill).

In closed loop recycling, the waste is converted to a raw material from which the same product may be re-manufactured. In open loop recycling, a product is manufactured for a secondary application from the recycled material, according to the limitations imposed by factors such as poor quality. Both methods are environmentally beneficial. The recycling process requires the input of energy and additional materials and emits various pollutants to air, water and soil. It is therefore less environmentally beneficial than reuse.

The third option is incineration that converts the waste to heat, ash and flue gas. This can be performed with or without energy recovery, although the latter is preferable in terms of environmental conservation. Incineration is generally not a preferred option because of the emissions and ash produced. For this reason, incineration is not accepted in many countries.

The final and least preferred option is disposal to landfill. Historically, this has been the most common method of waste disposal, as there was at one time no shortage of land for the disposal of waste. However, almost every country is now running out of landfill space and waste management options have become a priority. Pollution from landfill gas and leachate are also major environmental concerns.

Waste treatments consisting of any suitable combination of these four options have been modelled. A UK study on a T-shirt and blouse assumed the disposal phase to be incineration and indicated that 3 MJ of energy was conserved (−3 MJ) per each T-shirt and each blouse.[96] In a case study of polyester jackets and T-shirts, a combination of incineration and standard landfill was assumed. For the functional unit assumed, one T-shirt consumed 0.06 MJ and one polyester jacket consumed 0.14 MJ of energy.[53]

1.14 Summary: key challenges in assessing and reducing environmental impacts

This chapter has reviewed the textile supply chain and its detailed environmental implications. The complete life cycle model of textile products has been discussed and broken down into individual steps. Detailed descriptions of each life cycle phase of textile products, their operations, processes and environmental impacts are given in this chapter.

Owing to lack of data, it is difficult to compare different fibres, yarns and fabric manufacturing technologies, finished fabric and apparel production techniques within a common framework. No single study reported so far compares textile fibres in terms of their entire life cycle assessment or individual phases. The need for an overview of the environmental impact of different elements in the textile and clothing supply chain based on the life cycle assessment approach is evident. Such a study will be complex because of the unique circumstances of manufacturing locations in terms of the following aspects:

- energy sources;
- energy production and transmission;
- availability of natural and artificial energy sources for production;
- water sources;
- water production and transmission;
- availability of raw materials and other production inventories;
- transportation issues associated with the availability of resources;
- location of manufacturing and its proximity to ports, etc.;
- climatic conditions prevailing in the manufacturing region and their associated impacts;
- local regulations for emissions to air, water and land;
- treatment of waste in the locality.

The preparation of comparable international studies will therefore take some time and will require the cooperation of researchers across the world. Life cycle assessment in the clothing and textile sector is still in its early stages and more research is needed and many more textile products need to be studied and their impacts documented.

There is also a need for a detailed and regularly updated inventory database of the life cycle phases of textile materials for reliable secondary data sources, as well as for research purposes such as filling data gaps.

The difficulties involved, coupled with a decentralized sector in which the processes are intermittent, mean the assessment and reduction of environmental impacts in textiles and clothing supply chains are complex. The availability of reliable data poses a further difficulty in the life cycle assessment of textile products and requires the cooperation of all involved in the supply chain. The reduction of environmental impacts will necessitate changes in processes and products and will require innovation in current production practices, with associated cost implications.

1.15 Sources of further information and advice

- Chen HL, Burns LD. Environmental Analysis of Textile Products. Clothing and Textiles Research Journal. July 2006;24(3):248–261.
- Ecotextiles news website, www.ecotextile.com.
- 'Environmental Hazards of the Textile Industry', Environmental Update #24, published by the Hazardous Substance Research Centers/South and Southwest Outreach Program, June 2006;Business Week, 5 June 2005.
- OECOTEXTILES website, oecotextiles.wordpress.com/.
- Slater K. Environmental impact of textiles: production, processes and protection: Woodhead Publishing; 2003.
- Textile Industry Poses Environmental Hazards, OEcoTextiles, Available at: http://www.oecotextiles.com/PDF/textile_industry_hazards.pdf.

11.16 References

1. World Commission on Environment and Development. *Our Common Future*. Oxford: Oxford University Press; 1987:27. ISBN 019282080X.
2. Muthu SS, Li Y, Hu JY, Mok PY. Quantification of environmental impact and ecological sustainability for textile fibres. *Ecological Indicators*. 2012;13(1):66–74.
3. van Dam JEG. Environmental benefits of natural fibre production and use. In: *Proceedings of the Symposium on Natural Fibres, Rome, Italy*. October 20 , 2008:3–17. CFC technical paper 56.
4. EJF. *The Deadly Chemicals in Cotton*. London, UK: Environmental Justice Foundation in collaboration with Pesticide Action Network UK; 2007. ISBN No. 1-904523-10-2.
5. Information on organic cotton published in Patagonia's website. Available at: http://www.patagonia.com/us/patagonia.go?&assetid=2077, Accessed November 15, 2012.
6. US Environmental Protection Agency. *List of Chemicals Evaluated for Carcinogenic Potential*. 2001.
7. Laursen SE, Hansen J, Knudsen HH, Wenzel H, Larsen HF, Kristensen FM. *EDIPTEX: Environmental Assessment of Textiles*. Danish Environmental Protection Agency; 2007. Working Report 24.

8. National Agricultural Statistics Service Information, USDA, (Chapter 14), Agricultural Statistics Annual, 2011 statistics, Available at: http://www.nass.usda.gov/Publications/Ag_Statistics/2011/Chapter14.pdf.
9. Kramer SB, Reganold JP, Glover JD, Bohannan BJM, Mooney HA. Reduced nitrate leaching and enhanced denitrifier activity and efficiency in organically fertilized soils. *Proceedings of the National Academy of Sciences of the United States of America*. 2006; 103(12):4522–4527.
10. Tilman D, Cassman K, Matson P, Naylor R, Polasky S. *Nature*. 2002;418:71–77.
11. Cotton and the environment', Organic Trade Association, Available at: http://www.ota.com/organic/environment/cotton_environment.html, Accessed November 18, 2012.
12. Kadolph SJ, Langford AL. *Textiles*. 9th ed. Upper Saddle River, NJ: Prentice Hall; 2002.
13. US Department of Agriculture, 'Agricultural Chemical Usage: 2003 Field Crop Summary.'.
14. Patagonia's Website, 'Organic cotton', Available at: http://www.patagonia.com.au/about/technology/e-fibres/organic-cotton/, Accessed November 21, 2012.
15. Textile exchange's website, 'Benefits of organic cotton agriculture', Available at: http://farmhub.textileexchange.org/upload/learning%20zone/Benefits%20of%20organic%20cotton%20agriculture.pdf, Accessed November 21, 2012.
16. Kalliala EM, Nousiainen P. Environmental profile of cotton and polyester–cotton fabrics. *Autex Research Journal*. 1999;1(1):8–20.
17. Cherrett N, Barrett J, Clemett A, Chadwick M, Chadwick MJ. *Ecological Footprint and Water Analysis of Cotton Hemp and Polyester*. Stockholm, Sweden: Stockholm Environment Institute; 2005.
18. Barber A, Pellow G. *Life Cycle Assessment New Zealand Merino Industry, Merino Wool Total Energy use and Carbon Dioxide Emissions*. Pukekohe, Auckland, New Zealand: The Agri Business Group; 2006.
19. O Ecotextiles, 'Why is Recycled Polyester Considered a Sustainable Textile?', Available at: http://oecotextiles.wordpress.com/2009/07/14/why-is-recycled-polyester-considered-a-sustainable-textile/, Accessed November 21, 2012.
20. *The Life Cycle Assessment (LCA) of Organic Cotton Fiber*. Textile Exchange; November 2014 (Textile Exchange).
21. *Life Cycle Assessment of Cotton Cultivation Systems: Better Cotton, Conventional Cotton and Organic Cotton*. C & A Foundation; May 25, 2018.
22. González-García S, Hospido A, Moreira MT, Feijoo G. Life cycle environmental analysis of hemp production for non-wood pulp. In: *3rd International Conference on Life Cycle Management*. Zurich: University of Zurich at Irchel; 27–29 August 2007.
23. Van der Werf HMG, Van Geel WCA, Wijlhuizen M. Agronomic research on hemp (*Cannabis sativa L.*) in The Netherlands, 1987–1993. *Journal of the International Hemp Association*. 1995;2(1):14–17.
24. Bennett S, Snell R, Wright D. Effect of variety, seed rate and time of cutting on fibre yield of dew–retted hemp. *Industrial Crops and Products*. 2006;24:79–86.
25. Van der Werf HMG. Life cycle analysis on field production of fibre hemp, the effect of production practices on environmental impacts. *Euphytica*. 2004;140:13–23.
26. Van der Werf H, Turunen L. The environmental impacts of the production of hemp and flax textile yarn. *Industrial Crops and Products*. 2008;27:1–10.
27. Gonzalez-Garcia S, Hospido A, Feijoo G, Moreira MT. Life cycle assessment of raw materials for non–wood pulp mills: hemp and flax. *Resources, Conservation and Recycling*. 2010;54(11):923–930.
28. Cherrett N, Barrett J, Clemett A, Chadwick M, Chadwick MJ. *Ecological Footprint and Water Analysis of Cotton, Hemp and Polyester*. Stockholm, Sweden: Stockholm

Environment Institute report prepared for and reviewed by BioRegional Development Group and World Wide Fund for Nature (WWF Cymru); 2005.
29. Turunen L, Van der Werf HMG. *Life Cycle Analysis of Hemp Textile Yarn. Comparison of Three Hemp Fibre Processing Scenarios and a Flax Scenario*. Rennes, France: Institute National de la Recherche Agronomique (INRA) within the framework of the European Union HEMPSYS project; 2006.
30. Lloveras J, Santiveri F, Gorchs G. Hemp and flax biomass and fibre production and linseed yield in irrigated Mediterranean conditions. *Journal of Industrial Hemp*. 2006;11(1):3−15.
31. Gorchs G, Lloveras J. Current status of hemp production and transformation in Spain. *Journal of Industrial Hemp*. 2003;8(1):45−64.
32. Audsley E, et al. *Harmonisation of Environmental Life Cycle Assessment for Agriculture*. Final Report'. Concerted action AIR3-CT94-2028. European Commission DG VI Agriculture; 1997.
33. Brentrup F, Küsters J, Lammel J, Kuhlmann H. Methods to estimate on–field nitrogen emissions from crop production as an input to LCA studies in the agricultural sector. *International Journal of LCA*. 2000;5(6):349−357.
34. EMEP/CORINAIR. *Atmospheric Emission Inventory Guidebook*. Technical report No. 11. Copenhagen, Denmark: European Environment Agency; 2006.
35. 'Contribution à la lutte contre l'effet de serre. Stocker du carbone dans les sols agricoles de France?' Expertise Scientifique Collective. In: Arrouays D, Balcsdent J, Germon JC, Jayet PA, Soussana JF, Stengel P, eds. *Rapport d'expertise réalisé par INRA à la demande du Ministère de l'Ecologie et du Développement Durable*. Paris, France: INRA; 2002.
36. Carus M, Gahle C, Pendarovski C, et al. *Studie zur Markt- und Konkurrenzsituation bei Naturfasern und Naturfaser-Werkstoffen (Deutschland und EU)*. Hürth, Germany: Nova-Istitut GmbH; 2008.
37. Anthony Turner J. *Linseed Law: A Handbook for Growers and Advisers*. Hadleigh, Suffolk: BASF (UK) Limited; 1987. ISBN 0-9502752-2-0.
38. Ontario Ministry of Agriculture, Food and Rural Affairs (OMAFRA) Publication 811, 'Agronomy guide for field crops − Flax', Available at: http://www.omafra.gov.on.ca/english/crops/pub811/9flax.htm.
39. Flax and Hemp project, University of Wales, Bangor, Gwynedd, UK, 'Guidelines for growing flax', Available at: http://www.flaxandhemp.bangor.ac.uk/pdfs/GuidelinesForGrowingFlax.pdf.
40. L. Turunen and H. van der Werf (2006). Life Cycle Analysis of Hemp Textile Yarn, Comparison of Three Hemp Fibre Processing Scenarios and a Flax Scenario. French National Institute for Agronomy Research, France.
41. Dissanayake NPJ, Dhakal HN, Grove SM, Singh MM, Summerscales J. Optimisation of energy use in the production of flax fibre as reinforcement for composites. In: *International Conference on Flax and Other Bast Plants (Fiber Foundations − Transportation, Clothing and Shelter in the Bioeconomy)*. Canada: SASKATOON (Saskatchewan); 2008:47−58, 21−23 July 2008. Abstract 10, ISBN-13: 978-0-9809664-0-4.
42. Design for Sustainability, Quick Start Project, Sustainability Victoria 2012, Australia, Part 6, Textiles and the Environment', Available at: http://www.resourcesmart.vic.gov.au/documents/Quickstart6.pdf.
43. Barber A, Pellow G. LCA: New Zealand merino wool total energy use. In: *5th Australian Life Cycle Assessment Conference*. 22−24 November 2006 (Melbourne, Australia).
44. Biswas WK, Graham J, Kelly K, John MB. Global warming contributions from wheat, sheep meat and wool production in Victoria, Australia − a life cycle assessment. *Journal of Cleaner Production*. 2010;18:1386−1392.

45. Fletcher K. *Sustainable Fashion and Textiles: Design Journeys*. Earthscan Publications Ltd; 2008.
46. Soulivanh S. Environmental impacts of trade liberalization in the silk handicrafts sector of the Lao PDR. In: *Background Research Paper, Rapid Trade and Environment Assessment (RTEA)*. International Institute for Sustainable Development (IISD); 2007. Available at: http://www.iisd.org/pdf/2008/rtea_lao_silk.pdf.
47. Akter N, Rahman A, Chowdhury M. *Environmental Investigation and Evaluation of Sericulture Programme and Ayesha Abed Foundation*. Dhaka: BRAC, Research and Evaluation Division; March 1998.
48. *'The Global Fiber Market in 2011, Lenzing', Information Sourced from the Cellulose Gap*. Gherzi; February 2011. Available at: http://www.lenzing.com/en/concern/investor-center/equity-story/global-fiber-market.html.
49. Claudio L. Waste couture: environmental impact of the clothing industry. *Environmental Health Perspectives*. 2007;115(9):A449−A454.
50. *Polyester, Clean by Design*. Pukekohe, Auckland: Natural Resources Defense Council; August 2011. Available at: http://www.nrdc.org/international/cleanbydesign/files/CBD_FiberFacts_Polyester.pdf.
51. Franklin Associates Ltd. Life cycle analysis (LCA): Woman's knit polyester blouse. In: *Prepared for American Fibre Manufacturers Association*; 1993. Available at: http://www.fibersource.com/f-tutor/LCA-Page.htm.
52. Smith, Barker. Life cycle analysis of a polyester garment. *Resources, Conservation and Recycling*. 1995;14:233−249.
53. Steinberger JK, Friot D, Jolliet O, Erkman S. A spatially explicit life cycle inventory of the global textile chain. *International Journal of Life Cycle Assessment*. 2009;14(5): 443−455.
54. *Eco-profiles of the European Plastics Manufacturers*. Bottle Grade: Poly–ethyleneterephthalate (PET); April 2010.
55. Environment in Textile Supply, Polyester Production', Available at: http://www.eco-forum.dk/textile-purchase/index_files/Page1145.htm.
56. Nylon 6 and 66', OEcotextiles, Available at: http://oecotextiles.wordpress.com/2012/06/05/nylon-6-and-nylon-66/.
57. *Chemicals Released during Open Burning*; 12/12/2005. Available at: http://denr.sd.gov/des/wm/sw/documents/OpenBurningChemicalList.pdf.
58. Boustead I. *Eco-profiles of the European Plastics Manufacturers*. March 2005. Nylon 66.
59. Boustead I. *Eco-profiles of the European Plastics Manufacturers*. March 2005. Nylon 6.
60. Boustead I. *Eco-profiles of the European Plastics Manufacturers*. March 2005. Low Density Polyethylene.
61. Boustead I. *Eco-profiles of the European Plastics Manufacturers*. March 2005. High Density Polyethylene.
62. Boustead I. *Eco-profiles of the European Plastics Manufacturers*. March 2005. Polypropylene.
63. Horrocks AR, Hall ME, Roberts D. Environmental consequences of using flame–retardant textiles − a simple life cycle analytical model. *Fire and Materials*. 1997;21(5):229−234.
64. Breast cancer and acrylic fibers', OEcotextiles, Available at: http://oecotextiles.wordpress.com/tag/textiles/.
65. Acrylic Plastics', Madehow.com, Available at: http://www.madehow.com/Volume-2/Acrylic-Plastic.html.
66. Making Rayon Fibre 1999', Available at: http://www.eng.auburn.edu/~drmills/mans382/Rayon/Making%20Rayon%20Fiber.pdf.

67. Eucalyptus fiber by any other name', OEcotextiles, Available at: http://oecotextiles.wordpress.com/category/fibers/viscose/.
68. Clean by design, Viscose Rayon', Available at: http://www.nrdc.org/international/cleanbydesign/files/CBD_FiberFacts_ViscoseRayon.pdf.
69. Gelbke HP, Göen T, Mäurer M, Sulsky SI. A review of health effects of carbon disulfide in viscose industry and a proposal for an occupational exposure limit. *Critical Reviews in Toxicology*. 2009;39(Suppl. 2):1–126.
70. Walters A, Santillo D, Johnston P. *An Overview of Textiles Processing and Related Environmental Concerns*. UK: Greenpeace Research Laboratories; 2005. Technical Note: 08/2005.
71. Environmental Compendium, Kvadrat, Version 3, Available at: http://www.kvadrat.dk/fileadmin/user_upload/downloads/environment/Kvadrats_Environmental_Compendium_0907_v3.pdf.
72. Textile Industry Self-Monitoring Manual. *Ministry of State for Environmental Affairs (MSEA)*. Cairo, Egypt: Egyptian Environmental Affairs Agency (EEAA); 2003.
73. Textile Industry Inspection Manual. *Ministry of State for Environmental Affairs (MSEA)*. Cairo, Egypt: Egyptian Environmental Affairs Agency (EEAA); 2003.
74. Wakelyn PJ, Greenblatt GA, Brown DF, Tripp VW. Chemical properties of cotton dust. *American Industrial Hygiene Association Journal*. 1976;37(1):22–31.
75. Iyakaew N, Chawsithiwong B. Intention to use mask for cotton dust among workers of spinning and weaving operations in Rajburana District, Bangkok, Thailand. *European Journal of Social Sciences*. 2012;31(1):115–120.
76. Fishwick D, Fletcher AM, Pickering CA, McL Niven R, Faragher EB. Lung function in Lancashire cotton and man made fibre spinning mill operatives. *Occupational and Environmental Medicine*. 1996;53(91):46–50.
77. Jiang CQ, Lam TH, Kong C, et al. Byssinosis in guangzhou, China. *Occupational and Environmental Medicine*. 1995;52(4):268–272.
78. Siziya, S. and Munalula, B. ( ) Respiratory conditions among workers in a cotton spinning mill in Zambia. ATDF Journal, 2(3): 9–11.
79. Ma Q, Li D, Zhong Y. A prospective study on respiratory symptoms and functions in new employees exposed to cotton dust. *Chinese Journal of Preventive Medicine*. 1997;3196: 355–357.
80. Wang XR, Pan LD, Zhang HX, Sun BX, Dai HL, Christiani DC. Follow–up study of respiratory health of newly hired female cotton textile workers. *American Journal of Industrial Medicine*. 2002;41(2):111–118.
81. United States Department of Energy. *Energy use, Loss and Opportunities Analysis: U.S. Manufacturing & Mining*; 2004. Available at: https://www.eecbg.energy.gov/industry/intensiveprocesses/pdfs/energy_use_loss_opportunities_analysis.pdf.
82. Hasanbeigi A. *Energy-Efficiency Improvement Opportunities for the Textile Industry*. Environmental Energy Technologies Division; September 2010. Available at: http://eetd.lbl.gov/sites/all/files/publications/lbl-3970e-ee-textilesep2010_1.pdf.
83. Koç E, Kaplan E. An investigation on energy consumption in yarn production with special reference to ring Spinning. *Fibres and Textiles in Eastern Europe*. 2007;15(4):63.
84. BAT Guidance Note on Best Available Techniques for the Textile Processing Sector' (1st ed.), Environmental Protection Agency, Ireland, Available at: http://www.epa.ie/downloads/advice/bat/bat%20guidance%20note%20for%20textiles%20processing%20sector.pdf.
85. Collins M, Aumonier S. *Streamlined Life Cycle Assessment of Two Marks & Spencer plc Apparel Products*. Draft Final Report by Environmental Resources Management; 2002.

Retrieved from: https://images-na.ssl-images-amazon.com/images/G/02/00/00/00/24/28/67/24286782.pdf.
86. Kalliala E, Talvenmaa P. Environmental profile of textile wet processing in Finland. *Journal of Cleaner Production*. 2000;8:143–154.
87. Blackburn RS, Payne JD. Life cycle analysis of cotton towels: impact of domestic laundering and recommendations for extending periods between washing. *Green Chemistry*. 2004;6(7):G59.
88. Jelse K, Jenny W. *Life Cycle Assessment of Dunicel Table Cover and Alternative Products, Final Report*. Stockholm: Swedish Environmental Research Institute; August 2011.
89. Environment Agency. *Life Cycle Assessment of Disposable and Reusable Nappies in the UK*. May 2005.
90. Muthu SS, Li Y, Hu J, Mok PY, Liao X. Carbon footprint of production processes of polypropylene nonwoven shopping bags. *Fibres and Textiles in Eastern Europe*. 2012; 3(92):12–15, 20.
91. Muthu SS, Li Y, Hu JY, Mok PY. Carbon footprint of shopping (grocery) bags in China, Hong Kong, and India. *Atmospheric Environment*. 2011;45(2):469–475.
92. Abena, 'Life Cycle Assessment (LCA) of Absorbing Incontinence Products', Available at: http://bambonatureusa.com/files/7913/1804/6817/life-cycle-assessment.pdf.
93. WRAP UK. *Water and Chemical Use in the Textile Dyeing and Finishing Industry*; 1997. Available at: http://www.wrap.org.uk/sites/files/wrap/GG062.pdf.
94. Altun Sule. Life cycle assessment of clothing process. *Research Journal of Chemical Sciences*. 2012;2(2):87–89.
95. European Commission. *Environmental Improvements Potential of Textiles (IMPRO Textiles)*. France: Bio Intelligent Service; March 17 , 2011. Available at: http://www.keystone-group.co.uk/clothing/proceedings/Yannick%20LeGuern.pdf.
96. Allwood JM, Bocken N, Laursen SE, Malvido de Rodriguez C. *Well Dressed? The Present & Future Sustainability of Clothing & Textiles in the UK*. Cambridge: University of Cambridge, Sustainable Manufacturing Group, Institute for Manufacturing; 2006.
97. Collins M, Aumonier S. *Streamlined Life Cycle Assessment of Two Marks & Spencer Plc Apparel Products*. London: Environmental Resource Management; 2002.

Ways of measuring the environmental impact of textile processing: an overview

2

2.1 Introduction

Any processing or manufacturing sequence for a product is responsible for creating multiple environmental impacts and the manufacturing industry is one of the biggest starting points for various vulnerable impacts on our planet. One can point out numerous causes for the environmental impacts produced by manufacturing sectors; some of the earmarked root causes include

- over-consumption and over-production;
- environmental impacts pertaining to the extraction of raw materials;
- impacts due to the production of chemicals, other materials and auxiliaries for production;
- usage of energy from different sources and its extraction effects;
- usage of water from different sources for processing and cooling and its extraction effects;
- emissions to air, water and land;
- treatment and discharge of effluents;
- production of solid waste and its disposal;
- impacts pertaining to transportation of materials, semi-finished and finished goods.

This list is applicable to any manufacturing sector, including textiles particularly because its supply chain is lengthy, as discussed in Chapter 1. There are various tools to assess the environmental impacts of textile products and textile processing. There are also many key indicators to show the level of environmental impacts caused by textile processing.

The long supply chain in textiles is responsible for a considerable number of different types of detrimental environmental impacts. These include the consumption of huge amounts of raw materials, water, energy, chemicals and auxiliaries and also the emission of significant amounts of pollutants to the air, water and land. To identify these environmental impacts and control them, there is a range of legislation in many countries. There are also numerous environmental standards and schemes applicable to textile products.

2.2 Ways of measuring the environmental impacts of textile processing and textile products

Environmental impact assessment is one of the most important issues facing any country. Governments and manufacturers, as well as many members of the public, are deeply concerned about the detrimental environmental outcomes of producing a

Assessing the Environmental Impact of Textiles and the Clothing Supply Chain
https://doi.org/10.1016/B978-0-12-819783-7.00002-8

product by deploying various raw materials and other inputs. These predominately negative environmental consequences (the vulnerable impacts) cover various important issues such as emissions to air, water and land, solid waste production and solid waste disposal. All these impact both human health and the environment. The importance of these issues mean it is now mandatory for governments and businesses to consider them in policy decisions. Most manufacturers now need to have a certain amount of information on the environmental impact of their operations. This information is centered on the concept of sustainable development. It is not important to describe the concept and theory of sustainable development in this book, as plenty of information is already available on this topic elsewhere. Suffice it to say that sustainable development revolves around three pillars, namely, environment, society and economy. In this book, it is environmental sustainability that is discussed.

This environmental information is related to a product in its entire life cycle and in each step of its manufacturing sequence. There are two ways of measuring the environmental impacts of textile production: one is the manufacturing level called as manufacturing sustainability and the other is a holistic approach to measure the impacts across the entire life cycle phases of a textile product called as product sustainability. In terms of product sustainability, to evaluate the environmental impact of products and processes, as well as to disseminate the related information, a wide array of concepts and tools are available. These tools help the formation of operational methods and serve as the means for reasoning, analysis and communication.[1] Tools are characterized into two types: *procedural* and *analytical*. As the name implies, procedural tools focus on procedures to aid decision making, whilst analytical tools deal with technical phases.[1,2]

A number of important concepts related to environmental sustainability at product level are listed in the following, with some information describing each one.[1]

Life cycle thinking

This is a holistic approach involving evaluating the impacts of a product across its entire life cycle, including the extraction of raw materials, the production process, transportation, consumption and disposal. This approach primarily aims to reduce the resources being employed and the pollutants being emitted in a product's life cycle as well as to improve its social and socio-economic performance.[3]

Life cycle management

This refers to a product management system that focuses on reducing the environmental and socioeconomic impacts of a product in its complete life cycle and value chain. Life cycle management (LCM) enables companies to reduce their environmental impacts through the continuous improvement of the entire production process. LCM does not refer to a single tool or method but rather a complete management process that collects, systemizes and circulates all the information related to a product during its lifetime.[4]

Design for the environment

This is a concept created in the early 1990s by the United States Environmental Protection Agency (USEPA). It works on the core principle of finding out the optimal ways of preventing pollution, thereby protecting people and the environment from damage.[5] Design for the Environment (DfE) helps businesses to save costs, reduce business and environmental risk and also to expand business opportunities.[6] DfE earmarked the following six approaches as central in achieving its core principle:[6]

1. technology assessments
2. formulator approach
3. best practices approach
4. greening the supply chain
5. integrated environmental management systems
6. life cycle assessments.

Cleaner technology

The ultimate aim of clean technology is to reduce the environmental footprint and to minimize environmental pollution. There is no standard definition for clean technology and many agencies/authors have defined this in their own way. In general, it is an attempt to keep the environmental burden of technology at a bare minimum, more specifically in terms of modifications to process systems, production systems, product attributes and techniques. The following are the attributes clean technologies must gain in order to achieve their goals:[7]

- raw material conservation;
- optimal use of raw materials and other materials;
- optimal use of other valuable ingredients, namely, energy and water;
- optimization of production processes;
- safe disposal of unwanted waste;
- maximum possible recycling of unavoidable waste;
- prevention of accidents;
- prevention of pollution by risk management.

Eco-efficiency

The concept of eco-efficiency was introduced by the World Business Council for Sustainable Development (WBCSD) in the early 1990s. It is based on the concept of using fewer resources to generate more goods and services and decreasing the levels of waste and environmental pollution.[8] A popular definition of eco-efficiency by WBCSD is 'being achieved by the delivery of competitively priced goods and services that satisfy human needs and bring quality of life, while progressively reducing ecological impacts and resource intensity throughout the life cycle, to a level at least in line with the Earth's estimated carrying capacity'.[8,9] WBSCD earmarked seven aspects of eco-efficiency:[10]

1. reducing material intensity of goods and services,
2. reducing energy intensity of goods and services,
3. reducing the dispersion of any toxic materials,
4. enhancing the recyclability of materials,
5. making the maximum possible utilization of renewable resources,
6. enhancing the durability (shelf time) of products,
7. improving service intensity of goods and services.

Industrial ecology

The concept of industrial ecology (IE) focuses on the study of energy, material flows and transformations through industrial systems. IE is primarily intended to promote the principles of sustainable development, i.e. the sustainable use of resources, the preservation of ecological and human health and the promotion of environmental equity at local, regional and global levels.[11,12]

Following on from these concepts, a couple of well-known analytical tools based on physical metrics have become available. These include life cycle assessment (LCA), material flow analysis/substance flow analysis, material intensity per service unit (MIPS), cumulative energy requirements, environmental input/output analysis and environmental risk assessment.[1] A few of these are discussed briefly.

Life cycle assessment

This is the most important and well-known technique for assessing the overall environmental impact of a product, process or service. LCA is employed to assess these impacts of products from cradle to grave, encompassing various life cycle phases, which include raw material extraction, production, transportation, use and disposal. The International Standards Organization (ISO) has outlined the processes required to carry out an LCA study. The technique of LCA revolves around the acquisition and evaluation of quantitative data on the various inputs and outputs of relevant materials, energy flows and waste flows to analyse various possible environmental impacts (both beneficial and detrimental).

Material flow analysis/substance flow analysis

This refers to an analytical method of systematically analysing the flows and stocks of materials under well-defined conditions. This is one of the more important tools, finding its applications in a number of areas, including IE, environmental engineering, resources management and waste management. It can be applied on various spatial and temporal scales for a wide array of materials, substances, processes and goods.[13,14]

Material intensity per service unit

MIPS was developed by Germany's Wuppertal Institute in the 1990s. Being measured in kilograms per unit of service, MIPS attempts to specify the resource quantity used to manufacture a designated product or service. MIPS then classifies these material inputs into five categories: abiotic raw materials, biotic raw materials, earth movements in agriculture and silviculture, water and air.[15]

Listed in the following are some of the procedural tools used for environmental sustainability:[1]

- environmental management system (EMS)
- environmental audit
- environmental labelling
- eco-design
- environmental impact assessment
- green procurement
- total quality environmental management system
- environmental performance evaluation.

This section not only describes in detail the LCA concept and its variants as well as the concepts of carbon and ecological footprints but also enumerates the differences between these concepts.

2.2.1 Product sustainability life cycle assessment: a brief introduction

Details about the LCA concept, including its history, theory, methods, models and standards, will be presented in Chapter 6, but a brief outline about LCA is presented here.

LCA is the crux of eco-design, dealing with the design approach of a product with full consideration to the environmental impacts made by the product in its entire life cycle. The LCA analyses the effects on the environment by both the use of resources (inputs) and the emissions created by a given process (outputs).

Inputs

- raw materials
- water
- energy
- chemicals and other auxiliaries.

Outputs

- product
- co-product
- solid waste
- air emissions
- water emissions
- emissions to land.

This whole quantification process starts at the raw material production and extraction phase, then spans the manufacturing process, progressing later into packaging, distribution, retail, use and disposal phases. Figure. 2.1 depicts the various phases involved in the life cycle of a product considered for the quantification of LCA. The

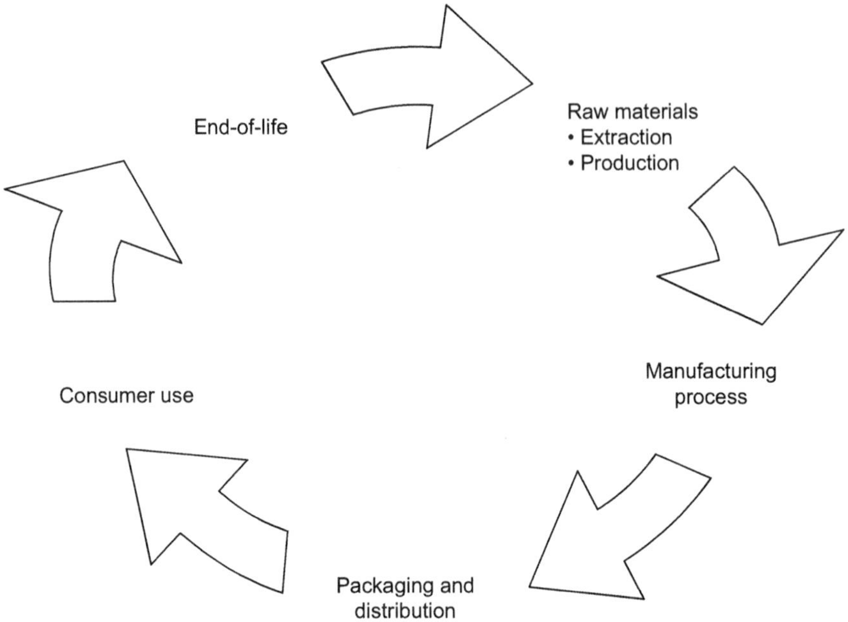

Figure 2.1 Various life cycle phases of a product.

analysis is not complete once the different inputs and outputs have been collected, this is only the initial step. These input and output details, termed as life cycle inventory, will be converted to mathematical models for analysis, a phase defined as 'impact assessment'. These impacts are then quantified in the LCA and related to a three-tier scale, namely, local, regional and global.

LCA can measure a long list of impacts on local, regional and global levels, for example,

- climate change (carbon footprint)
- ecological footprint
- water footprint
- acidification
- eutrophication
- human toxicity
- energy footprint
- ozone depletion potential
- photochemical oxidation potential
- smog
- depletion of biotic and abiotic resources
- eco-damage
- land use.

LCA can be conducted for a range of products, processes and services. It can be performed in two steps: a screening or preliminary assessment and a detailed or full-scale assessment. Details of these steps will be shown in Chapter 6.

LCA quantifications can be carried out in many forms, which are labelled as variants of LCA. Following are the most popular of these variants:

- Cradle to grave—a full life cycle assessment that includes all the stages of a life cycle.
- Cradle to gate—an LCA that deals only with the raw material extraction, production, manufacturing, packaging and transportation processes. It assesses only the activities that occur within the factory. It will not include the distribution, consumer use and disposal phases.
- Cradle to cradle – typically a cradle to grave assessment, where the end-of-life stage of a product is a recycling process, thereby the product will not be discarded after the end of life.

These variants, along with others, will be discussed in detail in Chapter 6.

2.2.2 Product sustainability: product carbon and ecological footprints

In this modern age, one of the most widely used environmental terms is *carbon footprint*. Owing to the alarming consequences of climate change, the measurement of a product or organization's carbon footprint has become so popular that it has been subsequently demanded by governments, consumers and other stakeholders. Currently, the assessment of a product's carbon footprint is being carried out by almost all industries, including the textile industry.

A carbon footprint is the measurement of the amount of greenhouse gases (CO_2, CH_4, N_2O, HFC, PFC and SF_6) emitted as a result, either directly or indirectly, of human activity. This is expressed in units of carbon dioxide emitted,[16] and carbon footprints can be assessed for both products and organizations. However, in this book, we deal primarily with the carbon footprint of products. A more in-depth study into the theory of the carbon footprint, the history of carbon footprint measurements and models of assessment will be presented in Chapter 3.

Another popular environmental term is *ecological footprint*. This means the quantification of natural resources required for a given activity. This includes the amount of biologically productive land and sea required to regenerate a resource, as well as to deal safely with the produced waste.[17] Detailed discussions on the development of the ecological footprint concept, the measurement of ecological footprints and their standards and models of assessment will be presented in Chapter 5.

In essence, carbon footprint is a subset of ecological footprint, the latter originally being a concept proposed in 1996 by Wackernagel and Rees.[17] Within the ecological footprint sphere, the measurement of carbon footprint refers to the amount of land required to absorb the quantity of CO_2 created by human activity.[18]

Both carbon footprints and ecological footprints attempt to measure, each in different ways, the pressure exerted by human beings on the living planet. The concept of ecological footprint helps measure the impact of various human activities on the biosphere, whereas carbon footprint measures the same but regarding atmosphere.

Ecological footprint is a useful way of understanding the human appropriation of the earth's regenerative capacity and can be applied to products, regions, countries and even the world as a whole. Ecological footprint is expressed in terms of the global hectares (gha) of bio-productive land.[19]

Carbon footprint is extremely useful in understanding the total amount of greenhouse gas emission caused directly or indirectly by human activity on earth. Like ecological footprint, it has a potentially broad application and can be calculated for products, nations, organizations, governments and individuals. Carbon footprint is expressed in terms of kilograms or grams carbon dioxide equivalents.[19]

Both these indicators deal with LCA through the quantification of specific impacts. These impacts are always relevant to a particular indicator. Furthermore, different indicators assess a product's potential environmental impact through its entire life cycle chain. So, for any indicator, the crux is LCA. With the aid of LCA, different indicators can therefore measure relevant impacts and then convert and express the results in suitable units.

These footprint indicators in LCA can be applied to textile products in order to trace all relevant impacts in their entire life cycle. This will help achieve each of the objectives of a product-based LCA, such as eco-product design and development. LCA can also be applied to the entire textiles and clothing supply chain to identify any problematic areas in terms of relevant impacts studied. These areas are called 'hot spots' and their identification will assist in minimizing negative effects.

2.2.3 Manufacturing sustainability

As manufacturing is one of the detrimental phases in a product's life cycle phases, which demands special attention in terms of measuring and controlling the impacts, manufacturing-level sustainability is perceived with significant focus. There are attempts developed in the textile sector to alleviate the environmental impacts in the manufacturing phase of textile production (manufacturing phase of a textile product), and there are many tools dedicated to measure the environmental impact in a textile product's manufacturing phase. There are many elements under the manufacturing scope and the key ones are

1. EMS
2. energy and greenhouse gas emissions
3. water consumption
4. wastewater/effluent treatment
5. emissions to air
6. hazardous waste management
7. hazardous chemical management

2.3 Environmental legislation relating to textiles

The environmental impacts associated with the production of materials in the textile and clothing supply chain were discussed in Chapter 1. Beginning with the cultivation or extraction of raw materials up until the disposal of clothes, the textile industry creates a multitude of environmental impacts. In each phase, from use to processing, to dyeing and printing, consumer products (and, in particular, textile materials) generate hugely damaging impacts on the environment. This process therefore deserves, in every region of the world, a great deal of attention in order to lessen these impacts. Indeed, in every country, there are now a number of legislations related to the textile industry and some of the more significant of these will be discussed here.

2.3.1 Legislation in Europe

There are many pieces of legislation available in Europe pertaining to the textiles and clothing sector. Sections relevant to the environmental issues of the clothing sector, such as the directives that deal with issues surrounding industrial emissions and waste management, are discussed in this part. Most of the environmental issues focus on specific chemicals and harmful substances used in manufacturing textile products. The following four pieces of European legislation that are related to textile industries[20] will be discussed in detail:

1. Integrated Pollution Prevention and Control (IPPC).
2. Emission Trading System (ETS).
3. Regulation on Registration, Evaluation, Authorisation and Restriction of Chemicals (REACH).
4. Other directives with relevance to the sector.

Integrated pollution prevention and control

This directive primarily aims at preventing, minimizing and eliminating pollution. This is achieved by cutting down the most significant industrial and agricultural emissions throughout the European Union, and for this reason, it is of great relevance to the textile industry.[20,21]

The original IPCC directive (Council Directive 96/61/EC of 24 September 1996 concerning IPPC) was implemented as Directive 96/61/EC and has been amended four times after it was put into force.[22] This directive demands integrating operating permits from the member states to control some of the specific industrial activities. This would be by means of considering the following aspects:[21,23]

- resource and energy efficiency;
- raw material and energy use;
- methods of operation of the site and the kind of technology used;
- emissions to water, air and soil;
- waste production management;
- prevention of accidents.

According to this directive, a plant is subjected to the IPCC directive if its capacity exceeds 10 tonnes per day, in regard to the operations of pre-treatment on wet processing, such as the bleaching, mercerization and dyeing of textiles. It is mandatory for such plants to satisfy the conditions in this permit in order to obtain an authorization to operate. These permit conditions have to be based on best available techniques (BAT).[20]

BAT is an important part of this directive and was indeed introduced by it. BAT is defined in Section 5 of the Environmental Protection Agency (EPA) Acts, 1992 and 2003, and Section 5(2)of the Waste Management Acts 1996 to 2005, as the 'most effective and advanced stage in the development of an activity and its methods of operation, which indicate the practical suitability of particular techniques for providing, in principle, the basis for emission limit values designed to prevent or eliminate or, where that is not practicable, generally to reduce an emission and its impact on the environment as a whole'. The definition of BAT is

B—'Best' implies the most effective way to achieve a high general level of protection of the environment as a whole.

A—'Available techniques' refer to those techniques developed on a scale that allows them to be implemented in the relevant activity under economically and technically viable conditions. This will take into consideration a number of factors, for instance, whether or not the techniques are used or produced inside the member state under discussion.

T—'Techniques' refer to both the technology used and the way in which the installation is designed, built, managed, maintained, operated and decommissioned.[24,25]

The reference document on BAT (BREF) is available for various industries. The BREF for the textile industry was published in July 2003 (BREF 07.2003). This deals with the installations involved in pre-treatment operations or the dyeing of textiles or fibres, with special attention given to the following textile operations:

- fibre preparation process;
- pre-treatment operations;
- dyeing process;
- printing process;
- finishing process.

In addition, upstream dyeing processes, which create major environmental impacts, are discussed. In this BREF, all important fibre types including natural fibres, regenerated cellulosic fibres, synthetic fibres and their blends are discussed.[26,27]

Emission trading system

Another important piece of legislation applicable to the textiles and clothing sector is the ETS. This is a central part of the European Union's policy of reducing greenhouse gas emissions from industrial sectors in a cost-effective and an economically efficient way. It corresponds to Directive 2009/29/EC published in 2009, which amended the previous Directive 2003/87/EC, so as to improve and extend the greenhouse gas emission allowance trading scheme to the community.[20,28]

The ETS scheme revolves around the cap and trade principle. According to this principle, there is a cap (limit) on the total amount of specific greenhouse gases that can be emitted by industrial sectors in general, as well as production facilities, power plants and other installations. Within this cap, factories receive certain allowances on greenhouse gas emissions and any factory can sell or purchase these from other factories if needed. At the end of each year, every company or factory has to submit sufficient allowances to cover all its emissions. By failing to do this, a company will incur a heavy financial penalty.

Any company or factory reducing its emissions below the given target accrues additional allowances and can use these to cover any extra emissions in the future, or trade them to other companies that are in need of more allowances. The crux of this entire scheme is to ensure that greenhouse gas emissions are trimmed down. As of now, the majority of allowances are given for free, but the number of allowances given will be gradually reduced over time in order to keep greenhouse gas emissions at the minimum targeted level. The directive specifies that any sector or sub-sector shall be given free emission allowances if it is exposed to a high risk of carbon leakage.

It is estimated that in 2020, greenhouse gas emissions will be 21% lower than those in 2005. The scheme operates in 30 countries, three of which (Iceland, Liechtenstein and Norway) are outside the European Union. The programme includes a number of major installations, such as combustion plants, power plants, oil refineries and major production sectors such as the paper and textile industry. It also considers the emission of nitrous oxide from specific operations.[29] Moreover, the ETS directive is applicable to any textile company that has combustion installations with a total rated thermal input of more than 20 MW.

Regulation on registration, evaluation, authorization and restriction of chemicals

Although REACH is not a new concept, it is applicable to the textile industry because of the high usage of chemicals in the textile production processes. The REACH directive (EC 1907/2006) published in 2006 was implemented on 1 June 2007 to improve the former legislative framework on chemicals of the European Union. It primarily deals with the production and use of chemicals and chemical substances and their potential impacts on the environment and human health. REACH was set up with seven important objectives that were developed in line with the overall framework of sustainable development.

1. Protection of human health and the environment.
2. Improved transparency.
3. Promotion of non-animal testing.
4. Prevention of fragmentation of the internal market.
5. Maintenance and enhancement of the competitiveness of the EU chemical industry.
6. Integration with international efforts.
7. Conformity with EU international obligations under the WTO.[30]

REACH intends primarily to improve and to protect human health and the environment through earlier and more efficient identification of chemical substances.[31,32] Moreover, this legislation also aims to augment innovation and competitiveness in the EU chemicals industry. REACH legislation demands producers and importers of chemicals and chemical substances gather a sufficient amount of information on the intrinsic properties of any substances used. It also demands producers and importers to register the information in a central database run by the European Chemicals Agency (ECHA).[32,33]

Other directives with relevance to the sector

This section deals with any directives pertaining to textile products or the textile sector other than those discussed earlier. This includes the Biocides Directive and voluntary approaches such as the European Eco-label for textile products.

Completion of the Biocides Directive is mandatory if biocides are added to textile products. The Biocidal Products Directive 98/8/EC covers 23 product types, included in four main groups:

- disinfectants and general biocidal products,
- preservatives,
- pest control,
- other biocidal products.[34]

The EU Eco-label is a voluntary approach that was started with the objective of helping European customers distinguish the quality of environmentally friendly products in the market.[35] The EU Eco-label for textiles is the official mark in the European Union for sufficiently eco-friendly products. This can be awarded to all types of textile products and accessories. An eco-label on a product guarantees the following criteria:[36]

- limited use of harmful substances to the environment,
- limited use of substances harmful to health,
- reduced air and water pollution,
- shrinkage resistance to drying and washing,
- colour fastness with washing, perspiration, wet rubbing, dry rubbing and exposure to light.

The ecological criteria that need to be met before an Eco-label award is given are classified into three parts:[36]

- textile fibre criteria,
- criteria on processes and chemicals,
- fitness for use.

Criteria for textile fibres

The criteria for textile fibres consist of four sub-sections.

1. **Fibre type:** As per this class, all fibre types can be used except mineral, glass, metal, carbon and other inorganic fibres. As far as fibre content is concerned, the criteria for a given fibre type need not be met if that particular fibre contributes to less than 5% of the total weight of the fibres in that product. These same rules also apply to recycled fibres.
2. **Limitation of toxic residues in fibres:** In this class, limitations of the total amount of toxic residue used in fibres are also set. The expectations for different fibres in terms of toxic residue are
   - acrylic: acrylonitrile <1.5 Mg/kg
   - cotton: residues of certain pesticides <0.05 ppm
   - elastane and polyurethane: no organotin compounds
   - greasy wool and other keratin fibres: limitations of certain pesticides
   - man-made cellulose: Adsorbable organic halides (AOX) <250 ppm

   - polyester: antinomy <260 ppm
   - polypropylene: no lead-based pigments.
3. **Reduction of air pollution during fibre process:** In this class, criteria for air-polluting substances for different fibres are described.
   - Acrylic: acrylonitrile <1 g/kg
   - Elastane and polyurethane: aromatic diisocyanates <5 Mg/kg
   - Man-made cellulose: S < 120 g/kg (filament) and 30 g/kg (staple)
   - Polyamide: N20 < 10 g/kg polyamide 6 and <50 g/kg polyamide 6.6
   - Polyester: volatile organic compounds (VOCs) < 1.2 g/kg.
4. **Reduction of water pollution during the fibre process:** In this type, criteria for the use of various water-polluting substances in different fibres are
   - flax and other bast fibres: chemical oxygen demand (COD)/total organic carbon from water reduction by at least 75% (hemp) and 95% (flax, other)
   - viscose: Zn < 0.3 g/kg
   - cupro: Cu < 0.1 ppm
   - greasy wool and other keratin fibres: COD < 60 g/kg, 75% reduction of COD, off-site treatment; if on-site treatment, COD < 5 g/kg, 6 < pH < 9 and T < 40 °C (17).

Criteria for manufacturing (processes and chemicals)

This label deals with the criteria for all stages in the production of a product, including fibre production. Limitation of the use of substances harmful to the environment (in particular, the aquatic environment) and to the health process is dealt with in this part. The criteria are

- 90% of carding and spinning oil, lubricants and finishes for primary spinning and 95% of sizing preparations, detergents, fabric softeners and weight complexing agents shall be sufficiently biodegradable or else shall be recycled;
- polycyclic aromatic hydrocarbons in mineral oils <3% by weight;
- no cerium compounds, halogenated carriers;
- no heavy metals and formaldehyde in stripping and depigmentation;
- no Alkylphenol ethoxylates (APEOs), Ditallow Dimethyl Ammonium Chloride (DTDMAC), distearyldimonium chloride (DSDMAC), dihardenedtallowdimethyl ammonium compounds (DHTDMAC), Ethylenediaminetetraacetic acid (EDTA), linear alkylbenzene sulphonate (LAS), Diethylenetriamine Pentaacetic Acid (DTPA), chrome mordant dyeing;
- chlorine agents are excluded for bleaching yarns, fabrics and end products;
- level of impurities in dyes (in ppm): Ag < 100, Ba < 100, Co < 500, Se < 20, Fe < 2500, As < 50, Cd < 20, Cr < 100, Cu < 250, Hg < 4, Ni < 200, Pb < 100, Sb < 50, Sn < 250, Zn < 1500, Mn < 1000;
- level of impurities in pigments (in ppm): As < 50, Cd < 50, Cr < 100, Hg < 25, Pb < 100, Sb < 250, Zn < 1000, Ba < 100, Se < 100;
- no chlorophenols, polychlorinated biphenyl (PCB) and organotin compounds during transportation or storage;
- no biocidal or biostatic products active during use phase;
- discharge to the water of metal complex dyes based on Cu, Cr or Ni: max 20% (cellulose dyeing), 7% (other dyeing process); after treatment: Cu < 75 Mg/kg (fibre, yarn, fabric), Cr < 50 Mg/kg, Ni < 75 Mg/kg;
- no azo dyes that cleave to a list of aromatic amines;

- no dyes classified as carcinogenic, mutagenic and toxic for reproduction according to Directive 67/548/EEC;
- no potentially sensitizing dyes if fastness to perspiration >4;
- printing pastes $<5\%$ VOCs, and no plastisol-based printing;
- formaldehyde <30 ppm for products in direct contact with the skin, 20 ppm for products for babies and young children and 75 ppm for others;
- COD from wet-processing <25 g/kg; if on-site treatment, $6 < pH < 9$ and $T < 40\ ^{\circ}C$;
- no flame retardants or finishing substances containing $>0.1\%$ of substances classified as carcinogenic, mutagenic and toxic for reproduction and dangerous for the environment according to Directive 67/548/EEC;
- shrink-resistant finishes only allowed for wool slivers and loose scouted wool;
- coatings, laminates and membranes: no plasticizers or solvents assigned a list of R-phases according to Directive 67/548/EEC.[36]

Fitness for use

This part sets the criteria for the textile product's performance and durability. According to this part, the following tests should be carried out either on dyed yarn or on the final product:

- Dimensional changes during washing and drying – 8% for knitted products, 8% for terry towelling, 6% for other woven products and 2% removable and washable curtain and furniture fabric.[36]
- Colour fastness with perspiration (acid, alkaline), washing, wet rubbing, dry rubbing, light with specified criteria described as follows:
  - The colour fastness to washing shall be at least level 3–4 for colour change and at least level 3–4 for staining.
  - The colour fastness to perspiration (acid and alkaline) shall be at least level 3–4 (colour change and staining).
  - The colour fastness to wet rubbing shall be at least level 2–3.
  - The colour fastness to dry rubbing shall be at least level 4.
  - For fabrics to be used for furniture/curtains/drapes, the colour fastness to light shall be at least level 5. For all other products the colour fastness to light shall be at least level 4.[37]

2.3.2 Legislation in the United States

The USEPA has also set laws and regulations for this sector. Textile manufacturing is covered under the manufacturing sector (NAICS 313). This deals with the industries involved in spinning natural and man-made fibres into yarns, which are later converted to fabrics by weaving or knitting and followed by dyeing and finishing. A dedicated webpage for the textile sector refers to different compliance details, various laws as well as regulations including the National Emission Standards for Hazardous Air Pollutants (NESHAP).[38]

This section discusses the major environmental laws applicable to the textile manufacturing sector in the United States.

The clean air act

It is the United States' federal law to regulate air pollution on a national level.[39] This law defines the responsibilities of the EPA to preserve and improve the nation's air quality and to prevent stratospheric ozone depletion.[40] This involves the regulation of air emissions from both mobile and stationary sources. This act authorizes the EPA to set up National Ambient Air Quality Standards (NAAQS) to preserve public health and also to control the emission of hazardous air pollutants (HAPs).[41] Section 112 of this act is dedicated to the emissions of HAPs. This is applicable to textiles, as the textile production chain generates a number of HAPs such as VOCs. This act also deals with preventing any accidental release of hazardous chemicals.[42]

The clean water act

Similar to the Clean Air Act, the Clean Water Act is a federal law designed to regulate and eliminate water pollution. This act focuses on regulating both the discharge of harmful pollutants (toxic substances) into the waters of the United States and the quality of surface water.[43,44] Under this act, the EPA has defined industries' wastewater standards as well as standards for contaminants in surface water.[44] The act controls the direct discharge of industrial pollutants to surface waters from any industrial operations or premises. The textile industry deals with a huge amount of pollutants, all of which need to be discharged, such as heavy metals and solvents. For this rcason, it is necessary for the textile industry to be regulated by this act in the United States.

The toxic substances control act

This environmental law attempts to regulate new or already existing chemicals from posing a risk to human health or to the environment. This act also controls the distribution and use of hazardous chemicals, and the EPA has established an inventory database of chemicals under the act (dealt with in Section 5). Standards defined by the act may be applied at any time. Chlorofluorocarbons (CFCs), lead, asbestos and polychlorinated biphenyls are the chemicals that are regulated by the act (Section 6 deals with this issue).[45] As the textile manufacturing sector inevitably uses various chemicals, this act is applicable to it.

Other important environmental laws that are also applicable to textiles are

- The Safe Drinking Water Act,
- The Emergency Planning and Community Right-To-Know Act,
- Comprehensive Environmental Response, Compensation, and Liability Act (CERCLA or Superfund).

2.4 Current environmental standards and schemes in the industry

The textiles and clothing sector is responsible, from the beginning until the end of its lengthy supply chain, for various environmental issues. Manufacturing clothing is prone to creating multiple environmental impacts in all its phases, from the cultivation or production of fibre until the disposal of the clothing products. Some of the environmental issues pertaining to the textiles and clothing sectors are

- usage of huge amounts of pesticides and synthetic fertilizers for the cultivation of natural fibres;
- deployment of huge amounts of water and energy for the production processes of textile products;
- consumption of non-renewable resources in different forms in the entire supply chain, especially for synthetic fibres;
- issues pertaining to the production of a huge amounts of waste and its disposal;
- emissions to air, water and land from the huge manufacturing link and their associated pollution issues;
- discharge issues associated with various pollutants;
- consumption of hazardous substances and chemicals in the production processes.

In view of these environmental issues created by the textiles sector, there is a pressing need to formulate various environmental standards and schemes for this sector to minimize its potential environmental impacts. There are a massive number of standards and schemes available in the world, and this section highlights the important environmental standards/schemes applicable to the textile industry.

2.4.1 *Global organic textile standard*

Recently, the popularity of organic textiles rose rapidly owing to a wide ranging increased demand of organic products. To cater for this, the organic textiles market was in dire need of a unique single standard to replace the current range of standards employed. In view of this, the Global Organic Textile Standard (GOTS) was developed to amalgamate organic standards within the textile industry. Because of this, manufacturers could sell their products to various countries under one certification umbrella. GOTS was developed by the International Working Group, members of which include the Organic Trade Association (OTA), the Soil Association (UK), the International Association of Natural Textile Industry (IVN) (Germany) and the Japan Organic Cotton Association.

GOTS is one of the most important and pioneering textile processing standards for organic textile manufacturers in the world. Applicable to all textile products made from organic fibres, GOTS demands manufacturers follow specific environmental criteria throughout the entire supply chain, and also requires them to comply with certain social criteria. To become GOTS certified, a textile product must be made out of at least 70% organic fibres.[46]

GOTS consists of two individual and indispensable parts: the GOTS standard and the GOTS certification. The GOTS standard is applicable to the production process, manufacturing sequences, packaging entities, transportation, labelling and the export and import of organic textile products. It covers the entire production and processing sequence down the supply chain of an organic textile product. The GOTS certification can be obtained for organic textile products of any form but is mainly awarded to yarns, fabrics, finished products (apparels), home textiles and even to fibre stage products.[2]

GOTS enlisted various mandatory elements for a textile product to be eligible for certification. Among these, a few key points are

- fibre content should be 70% of organic fibres and requiring social criteria;[46]
- all chemical inputs (dyestuffs and auxiliaries, etc.) have to meet specified environmental and toxicological requirements[46,47]; the choice of accessories is limited in accordance with ecological aspects also[46,47];
- it has to comply with the specific requirements laid down for fibre production, processing and test parameters for various production steps dictated in the recent version 5.0 of GOTS, which was released on March 2017 (the very latest one, which is version 6.0, is under preparation now);[47]
- a wastewater treatment plant with proper functioning is mandatory for a chemical processing plant[46]; all processors must comply with social criteria conditions as well as with those laid in the standard[46];
- a long list of prohibited and restricted inputs to be followed as well, such as flame retardants, chlorinated benzenes, short-chain chlorinated paraffins, and so on.[47]

GOTS certification has to be obtained from the authorized (accredited) certification bodies of GOTS after thorough on-site inspection, verification and confirmation of GOTS standard and procedures.

2.4.2 Bluesign

The textile production sequence consumes huge amounts of chemicals and demands a large amount of resources. As per the details indicated by Bluesign, 1 kg of natural textile needs 350–1500 g of chemicals, whereas 1 kg of synthetic textile requires 110–820 g. 700 liters of water are consumed to finish 1 kg of textile.[48] Textile processing pollutes the environment badly and demands a great deal of ecological resources; therefore, there is a serious need for sustainable textiles, which significantly reduce both their consumption of resources and their emission of polluting substances. However, to achieve all this, it is necessary to have an effective evaluative standard.

Bluesign is an independent environmental textile standard developed to cope with the rising demand of sustainable and eco-friendly textile products. Tackling the environmental issues at their roots, the Bluesign standard works on the following five principles:[49]

1. **Resource productivity—cautious use of energy and raw materials:** Employing eco-efficient products to achieve superior performance with meaningfully reduced consumption of resources and minimum possible pollution to air and water, as well as a reduced ecological footprint.

2. **Air emissions—reduced air pollution to combat climate change:** The removal of harmful air emissions from all phases. Also enforces controlled emission limits through the complete supply chain. Furthermore, attempts to optimize the energy needs, which in turn reduces the amount of CO_2 emitted.
3. **Occupational health and safety—ensuring health and safety at work:** Protection of employees from exposure to hazardous materials by providing training to them on the handling and safe storage of chemicals.
4. **Water emissions—protection of the entire water cycle:** Controlling emissions of hazardous substances in water and ensuring the proper functioning of wastewater treatment plants.
5. **Consumer safety—ensuring responsibility for people and the environment:** Ensuring the use of safer components, processing conditions and procedures throughout the processes in the supply chain to reduce any risks to human beings and the environment.

The adoption of a Bluesign standard also ensures the prevention of hazardous substances from ever entering the supply chain. This involves meeting the prescribed threshold levels of all chemicals, accessories and even finished textile products. In addition to this, the Bluesign standard helps its customers to apply and implement their standard. This is done by providing them with the following three innovative tools:[49]

1. Bluetool—a web-based tool that helps chemical suppliers understand whether their products meet the Bluesign standard, as well as provides a list of all the important instructions for the evaluation of chemicals.
2. Bluefinder—a search engine to help clients select Bluesign-approved input streams, such as chemical substances. This is always kept up to date.
3. Blueguide—a web-based guide to Bluesign-approved fabrics and accessories.

The Bluesign standard aims to link manufacturers, retailers, suppliers, buyers and brands together to achieve a sustainable and profitable textile industry.[49]

2.4.3 Oeko-Tex standards

The Oeko-Tex standard focuses mainly on textile ecology. One of the popular certification systems of Oeko-Tex is Standard 100, which is an independent testing and certification scheme for raw, semi-finished and finished textile products at all processing levels, as well as accessory materials used.[50]

It was developed in 1992 to prevent the usage of certain chemicals.[50] It concentrates on the use of dangerous substances such as heavy metals, formaldehyde, toxic dyes and pesticides. Because the Oeko-Tex standard has developed such stringent requirements for the final product, certified products demonstrate that they are free from the substances that harm human health and the environment.

Textile products of any form are eligible for this certification, including raw and finished yarns, fabrics, apparels and home textiles. Owing to the nature of these products, namely, their being in contact with skin to different degrees, healthcare parameters for harmful chemicals will vary according to four classes:[50]

1. Class I—textile products for babies and toddlers up to 3 years.
2. Class II—textile products close to the skin such as underwear, T-shirts.
3. Class III—textile products worn away from the skin, such as jackets.
4. Class IV—furnishing textile products such as table cloths.

Oeko-Tex has developed a range of new services, which are listed in the following, along with the descriptions of each service defined by Oeko-Tex.[51]

- MADE IN GREEN: It is a traceable product label for textile products made of materials tested for harmful substances that, in addition, have been manufactured in environmentally friendly facilities and under socially responsible working conditions. The product ID and/or the QR Code on the label allows consumers to transparently trace labelled products.
- STeP—This stands for Sustainable Textile Production and is a certification system for environmentally and socially responsible production facilities along the textile chain. The aim of STeP is to continually improve production conditions and health and safety in all business areas.
- ECO PASSPORT—This is a certification system for textile and leather chemicals, colorants and auxiliaries. It consists of a three-step verification procedure that analyses whether or not substances meet specific criteria for a sustainable textile production.
- DETOX TO ZERO: This enables manufacturers along the textile chain based on an annual status report to assess whether their chemical management and the quality of their wastewater and sludge is in line with the goals of the Detox campaign.
- LEATHER STANDARD: This is a certification system for leather articles, at all levels of production, tested for harmful substances. The tested article receives the certificate if it complies with the annually updated Oeko-Tex requirements.
- MySTeP: It is a database that allows to monitor and assess the sustainability scores of suppliers along the textile value chain, while transparently managing existing Oeko-Tex certificates.

2.4.4 Eco-labels

An eco-label is an official label or mark given to help consumers to distinguish environmentally friendly products from those that are not. It is mostly applicable to food products and consumer goods. This eco-labelling scheme was started in the 1980s by various national labelling systems in different countries across the world to assist customers who are in need of choosing genuinely green products.

The crux of the eco-label scheme is embedded into the concept of LCA. This means evaluating the environmental impacts of a particular product through its entire life from cradle to grave. Any labelled product must prove to have lowered its environmental impact in all its life cycle phases. To streamline the labelling system, ISO has earmarked a series of standards (ISO 14020/21/22/23/24) for environmental labelling. ISO has classified eco-labels into three categories:[52,53]

- **Type I label— Eco-label (ISO 14024):** Voluntary, multi-criteria label based on life cycle considerations. This is awarded by a third party.
- **Type II label— Self-declared environmental claims (ISO 14021):** Product declaration by companies. This is not verified by an outside party and therefore runs the risk of being ineffective.
- **Type III label— Environmental declarations or environmental impact labels (ISO 14025):** These claims are based on quantified information of a product's environmental impact during its life cycle.

These eco-labels are quite familiar in the textile sector. Indeed, there are a large number of eco-labels (over 100) a textile product can bear, depending on the country of relevance.[54] The EU eco-label flower, a European voluntary label, is one of the most well-known eco-labels in the textile sector and it is applicable to many categories of products.[55,56] This label was discussed in detail in the previous section.

2.5 Summary: key methods reviewed in this book

Beginning with the cultivation of natural fibres or the production of synthetic fibres until the disposal of natural or synthetic materials, the textiles and clothing sector involves an enormous manufacturing link that consists of numerous intermittent and continuous processes. It can therefore have a massive environmental impact. In this chapter, a brief note on root causes of this on a local, regional and global scale was provided. Details pertaining to the various concepts of environmental impact, a number of different product assessments, as well as a range of processing sequences, were discussed. These included life cycle thinking, DfE, LCM, IE, cleaner production techniques and eco-efficiency. In addition, different tools for environmental impact assessment of products and processes, namely, procedural and analytical tools, were discussed. One of the most widely used and familiar analytical tools, LCA, was discussed in detail.

A brief outline of the concept of LCA as well as its most popular variants, namely, cradle to gate and cradle to grave assessments, were examined in this part. In addition to LCA, there are many environmental indicators currently gaining huge popularity, for instance, the concept of a carbon and ecological footprint. Details pertaining to these indicators, along with the differences between them and the LCA, were also discussed.

Along with the concepts and tools for the quantification of the environmental impact of textiles, this chapter presented the important aspects of a number of environmental laws and regulations applicable to the clothing sector. Environmental legislations applicable in the European Union were discussed in detail, including the IPPC Act, the ETS and REACH. The EU eco-label was discussed in detail, as well as other directives applicable to the textile sector. Various laws and regulations stipulated in the United States were examined in this chapter, including the Clean Air Act, the Water Act and the Toxic Substances Controlling Act.

In this chapter, an overview of various environmental schemes and standards applicable to textiles and clothing sector was also included. Key standards and schemes, namely, GOTS, Bluesign, Oeko-Tex and eco-labels, were elaborated.

2.6 Sources of further information and advice

- Baumann K, Cowell S. An Evaluative Framework for Environmental Management Approaches. Paper III, AFR report 183, Life Cycle Assessment and Decision Making—theories and practices. Technical Environmental Planning, Chalmers University of Technology, Göteborg; 1998.
- Bide M. Textiles and the environment. *Industrial Fabric Products Review*. 2001;78(1): 50–53.
- Fletcher KT. Design, the environment and textiles: developing strategies for environmental impact reduction. *Journal of The Textile Institute*. 1998;89(3):72–80.

- Joshi M. Environmental management systems for the textile industry: a case study. *Indian Journal of Fibre and Textile Research.* 2001;26(1):33–38.
- Nieminen-Kalliala E. Environmental indicators of textile products for ISO (Type III) environmental product declaration. *AUTEX Research Journal.* 2003;4. http://www.autexrj.org/No4–2003.html.
- SETAC-Europe Working Group on Conceptually Related Programmes. Life Cycle Assessment and Conceptually Related Programmes. Brussels: SETAC-Europe; 1997.
- Simpson JA, Binkley J, McMahon PH. New environmental legislation affecting textile-industry manufacturing. *Journal of The Textile Institute.* 2000;91(1):65–72.
- Tobler M, Scheidegger Y. Ecological key figures for benchmarking in textiles. *Fiber Society Conference, Session on Sustainability and Recycling, 23–25 May, St. Gall, Switzerland, 2005.*
- Udo de Haes HA, Jolliet O, Finnveden G, Hauschild M, Krewitt W, Müller-Wenk R. Best available practice regarding impact categories and category indicators in life cycle impact assessment. Background document for the second working group on life cycle impact assessment of SETAC-Europe. Parts 1 and 2. *International Journal of Life Cycle Assessment.* 1999;4:66–74 and 167–74.
- UNEP Industry and Environment. *Life Cycle Assessment: what it is and how to do it.* United Nations Publication Sales no. 9C-III-D.2. Paris; 1996.

2.7 References

1. Wrisberg N, Udo de Haes HA, Triebswetter U, Eder P, Clift R. *Analytical Tools for Environmental Design and Management in a Systems Perspective.* Dordrecht: Kluwer Academic Publishers; 2002.
2. Finnveden G, Moberg A. Environmental systems analysis tools – an overview. *Journal of Cleaner Production.* 2005;13:1165–1173.
3. A Special Publication for the RIO + 20 Summit and Beyond. *Greening the Economy through Life Cycle Thinking – Ten Years of the UNEP/SETAC Life Cycle Initiative.* United Nations Environment Programme; 2012.
4. *'Life Cycle Management – A Business Guide to Sustainability'.* United Nations Environment Programme; 2007.
5. Design for the Environment, USEPA's Website, Available at: http://www.epa.gov/dfe/pubs/about/index.html.
6. The US Environmental Protection Agency's Design for the Environment Program, US EPA's DFE Brochure, Available at: http://www.epa.gov/dfe/pubs/tools/DfEBrochure.pdf.
7. European Commission, Directorate General Environment. *LIFE FOCUS/Industrial Pollution, European Solutions: Clean Technologies, LIFE and the Directive on Integrated Pollution Prevention and Control (IPPC Directive).* Luxembourg: Office for Official Publications of the European Communities; 2003. ISBN 92–894-6020–2.
8. Wu Y, Yu D, Lee J. *Eco-efficiency Indicators: Measuring Resource-use Efficiency and the Impact of Economic Activities on the Environment.* United Nations Economic and Social Commission for Asia and the Pacific, United Nations publication; 2009. Available at http://www.neaspec.org/publication/Eco-efficiencyIndicators.pdf.
9. World Business Council for Sustainable Development. *Eco-Efficient Leadership for Improved Economic and Environmental Performance.* 1996. Geneva, Switzerland.
10. Lovins LH. Rethinking Production. In: *State of the World 2008.* 2008:34.

11. Frosch RA. Industrial ecology: a philosophical introduction. *Proceedings of the National Academy of Sciences USA*. 1992;89:800–803.
12. Garner A, Keoleian GA. *Industrial Ecology: An Introduction, Pollution Prevention and Industrial Ecology*. US: National Pollution Prevention Center for Higher Education, University of Michigan; November 1995.
13. Haberl H, Fischer-Kowalski M, Krausmann F, Weisz H, Winiwarter V. Progress towards sustainability? What the conceptual framework of material and energy flow accounting (MEFA) can offer. *Land Use Policy*. 2004;21(3):199–213.
14. Brunner PH, Rechberger H. *Practical Handbook of Material Flow Analysis*. Boca Raton, Florida: CRC Press LLC; 2004.
15. Ritthoff M, Rohn H, Liedtke C. *Calculating MIPS*. Wuppertal Institute of Climate, Environment and Energy; 2002.
16. Hertwich EG, Peters GP. Carbon footprint of nations: a global, trade-linked analysis. *Environmental Science and Technology*. 2009;43(16):6414–6420.
17. Wackernagel M, Rees W. *Our Ecological Footprint*. Gabriola Island, BC: New Society Publishers; 1996.
18. Pandey D, Agrawal M, Pandey JS. Carbon footprint: current methods of estimation. *Environmental Monitoring and Assessment*. 2011;178(1–4):135–160.
19. Galli A, Wiedmann T, Ercin E, Knoblauch D, Ewing B, Giljum S. Integrating ecological, carbon and water footprint into a 'footprint family' of indicators: definition and role in tracking human pressure on the planet. *Ecological Indicators*. 2011;16:100–112.
20. 'Textiles and clothing: environmental issues', European Commission, Available at: http://ec.europa.eu/enterprise/sectors/textiles/environment/index_en.htm#h2–1.
21. 'Integrated Pollution Prevention and Control Directive', Environment Agency, Available at: http://www.environment-agency.gov.uk/business/regulation/109813.aspx.
22. 'The IPPC Directive, Legislation', Available at: http://ec.europa.eu/environment/air/pollutants/stationary/ippc/legis.html.
23. European Commission, DG Environment-G2, B-1049, Bruxelles. *European Commission Green Public Procurement (GPP) Training Toolkit – Module 3: Purchasing Recommendations, Textiles-Background Product Report, Toolkit Developed for the European Commission by ICLEI – Local Governments for Sustainability*. 2008.
24. 'Best Available Techniques Guidance Notes', Environment Protection Agency, Available at: http://www.epa.ie/whatwedo/advice/bat/.
25. Eur-Lex, Access to European Law, 32008L0001, Directive 2008/1/EC of the European Parliament and of the Council of 15 January 2008 Concerning Integrated Pollution Prevention and Control (Codified Version) (Text with EEA Relevance), Available at: http://eur-lex.europa.eu/LexUriServ/LexUriServ.do?uri=CELEX:32008L0001:en:NOT.
26. *'Integrated Pollution Prevention and Control (IPPC)', Reference Document on Best Available Techniques for the Textiles Industry*; July 2003. European Commission, Available at http://eippcb.jrc.ec.europa.eu/reference/BREF/txt_bref_0703.pdf.
27. BREF for Textiles Industry', Available at: http://eippcb.jrc.es/reference/txt.html.
28. Directive 2009/29/EC of the European Parliament and of the Council of 23 April 2009, 5.6.2009, Available at: http://eur-lex.europa.eu/LexUriServ/LexUriServ.do?uri=OJ:L:2009:140:0063:0087:EN:PDF.
29. 'Climate Action'. European Commission, Emissions Trading System (EU ETS), Available at: http://ec.europa.eu/clima/policies/ets/index_en.html.
30. *'REACH in Brief'. European Commission, Environment Directorate General*; October 2007. Available at http://ec.europa.eu/environment/chemicals/reach/pdf/2007_02_reach_in_brief.pdf.

31. 'Chemicals, REACH', European Commission, Available at: http://ec.europa.eu/enterprise/sectors/chemicals/reach/index_en.htm.
32. REACH, European Commission', Available at: http://ec.europa.eu/environment/chemicals/reach/reach_intro.html.
33. Environment fact sheet: 'REACH', European Commission, Available at: http://ec.europa.eu/environment/pubs/pdf/factsheets/reach.pdf.
34. Biocides Directive, European Commission, Available at: http://ec.europa.eu/environment/biocides/main_subjects.html.
35. 'The Ecolabel Catalogue', European Commission, Available at: http://ec.europa.eu/ecat/.
36. 'The EU Eco Label for Textile Products', Available at: http://ec.europa.eu/environment/ecolabel/documents/factsheet_textiles.pdf.
37. Official Journal of the European Union, Commission Decision of 9 July 2009, Establishing the Ecological Criteria for the Award of the Community Eco-Label for Textile Products, Notified Under Document Number C(2009) 4595, L 197/70, Available at: http://eur-lex.europa.eu/LexUriServ/LexUriServ.do?uri=OJ:L:2009:197:0070:0086:EN:PDF.
38. Website of US EPA, 'Information on Textile Manufacturing Sector (NAICS 313)', Available at: http://www.epa.gov/lawsregs/sectors/textiles.html.
39. The Encyclopedia of Earth, 'Clean Air Act', United States, Available at: http://www.eoearth.org/view/article/151129/.
40. 'Clean Air Act', US EPA's website, Available at: http://www.epa.gov/air/caa/.
41. Summary of the Clean Air Act, US EPA, Available at: http://www.epa.gov/regulations/laws/caa.html.
42. 'RCRA in focus, Textile Manufacturing', US EPA, Available at: http://www.epa.gov/osw/inforesources/pubs/infocus/k02028.pdf.
43. Encyclopedia of Earth, 'Clean Water Act', United States, Available at: http://www.eoearth.org/view/article/51cbed3c7896bb431f690bc0/.
44. 'Summary of the Clean Water Act', US EPA, Available at: http://www.epa.gov/regulations/laws/cwa.html.
45. 'Toxic Substances Control Act (TSCA)', US EPA, Available at: http://www.epa.gov/oecaagct/lsca.html.
46. 'GOTS, The Standard', information from the GOTS website, available at: http://www.global-standard.org/the-standard.htm.
47. *Global Organic Textile Standard (GOTS) for Soil Association Certification Ltd Licensees, GOTS Version 5.0*. March 1, 2017.
48. *The Bluesign Standard, the Holistic Approach for Brands Textile Manufacturer*. November 2010.
49. Bluesign Standard Brochure, Bluesign website, Available at: https://www.bluesign.com/downloads/criteria/bluesign_system_v1_0.pdf.
50. Okeo-Tex Official website, Available at: https://www.oeko-tex.com/en/business/certifications_and_services/ots_100/ots_100_start.xhtml.
51. https://www.oeko-tex.com/en/business/certifications_and_services/certifications_and_services.xhtml.
52. 'A Guide to Environmental Labels – for Procurement Practitioners of the United Nations System, Executive Summary', UNEP, Available at: http://www.greeningtheblue.org/sites/default/files/EnvLabels_executivesummary_0.pdf.
53. Allison C, Carter A. *Study on Different Types of Environmental Labelling (ISO Type II and III Labels): Proposal for an Environmental Labelling Strategy*. DG Environment, European Commission; September 2000. Final Report.

54. All eco-labels on textiles, 'Ecolabel Index', Available at: http://www.ecolabelindex.com/ecolabels/?st=category,textiles.
55. https://www.oeko-tex.com/media/downloads/Factsheet_OEKO-TEX_Product_Portfolio_EN.pdf.
56. EU eco-label 'Flower', Eco-Institut website, Available at: http://www.eco-institut.de/en/from-analysis-to-quality-assurance/national-marks-of-conformity/eu-ecolabel-flower/.

Textile processing and greenhouse gas emissions: methods for calculating the product carbon footprint of textile products

3

3.1 Introduction

Amongst the various measures of environmental impacts affecting our planet, the concept of 'carbon footprint' is considered to be one of the most significant. A carbon footprint is a way of assessing the impact of an activity on the environment, specifically the production of gases such as carbon dioxide (the so-called 'greenhouse' gases [GHGs]) that have been linked to climate change. The concept has been discussed widely by the media, non-governmental organizations, governments, manufacturers, companies and the general public. Customers are increasingly demanding low-carbon products, and the communication of carbon footprint results has become one of the top priorities of many organizations. Governments in many countries have started to make disclosure of the carbon footprint of a product or process an essential requirement for many industries. Carbon trading has also become an important business sector. Various terms associated with reducing carbon footprints have become popular, such as clean development mechanism, carbon neutral, carbon reduction and low carbon.

The importance of the measurement and reduction of the emission of GHGs (i.e. carbon footprint) is understood by all the major industrial players. The textile industry, with its massive supply chain, is one of the largest contributors to GHG emissions. Beginning with the raw material extraction, textile production emits huge amounts of GHGs throughout the supply chain, and textile products are similarly responsible throughout their life cycle phases.

3.2 The main principles of carbon footprint measurement

Any introduction of carbon footprint must begin with the rise in global average temperatures by defining basic terms such as 'global warming' and 'greenhouse effect'. As these terms have been discussed thoroughly in other publications, this section will touch upon them only briefly.

Global warming refers to the spike in global average temperature near the Earth's surface and is principally caused by enhanced concentrations of GHGs in the atmosphere.

Assessing the Environmental Impact of Textiles and the Clothing Supply Chain
https://doi.org/10.1016/B978-0-12-819783-7.00003-X

On average, the Earth's temperature has increased by 1.4 °F over the past century and is expected to go up another 2–11.5 °F over the next 100 years.[1] The concentration of these GHGs has increased very significantly after industrialization, resulting in global warming. In its fourth assessment report, the Intergovernmental Panel on Climate Change (IPCC) has strongly recommended limiting the rise in global average temperature to below 2 °C similar to the pre-industrial level and this has encouraged the growing drive to combat global warming.[2,3] Global warming causes massive alterations in climatic patterns, which are evident from the current changes in weather and climate.[1]

The Kyoto Protocol, an international agreement on climate change, has identified six important gases:[4]

1. carbon dioxide (CO_2)
2. methane (CH_4)
3. nitrous oxide (N_2O)
4. hydrofluorocarbons (HFCs)
5. perfluorocarbons (PFCs)
6. sulphur hexafluoride (SF_6)

In this list, CO_2 is considered to be the most significant contributor to climate change issues. Each above-mentioned gas has its own global warming potential (GWP) based on its radioactive capacity compared to CO_2. Methane and nitrous oxide are the next most important gases, contributing considerably to the issue of global warming.[5]

3.2.1 Global warming potential

Each GHG possesses a unique heat trapping power and lifetime in the atmosphere. Having taken CO_2 as the reference point of GHGs, GWP is a tool to compare the potential of various gases in terms of their heat trapping capacity in the atmosphere with that of CO_2 over a specified period. The Kyoto Protocol identified 100 years as the time frame for the calculation of GWP values. As the reference gas, a GWP of 1 is assigned to CO_2 and it serves as the baseline unit. Each GHG has a unique GWP value, and these GWP values are very useful for calculating and converting various GHG emissions to comparable carbon dioxide equivalents.[6–8]

For a 100-year time horizon, GWP values of these six important GHGs provided by the IPCC in its fourth report (published in 2007) are carbon dioxide (CO_2)1; methane (CH_4)25; nitrous oxide (N_2O)298; HFCs in the range 124–14,800; PFCs in the range 7390–12,200; and sulphur hexafluoride (SF_6) 22,800.[8]

3.2.2 The concept of carbon footprint

The term 'carbon footprint' does not have its own definition as such but came from the concept of the ecological footprint, proposed in the 1990s by Rees and Wackernagel.[9,10] The idea and measurement of a carbon footprint is embedded into the measurement of the ecological footprint as well as being a separate entity. Under the umbrella of the ecological footprint measurement, carbon footprint helps quantify the amount of land needed to absorb the total amount of CO_2 produced by mankind over its lifetime. Since the issue of global warming came into existence and has started gaining worldwide attention, the carbon footprint was considered separately and is now one of the most important ecological factors that has to be measured and accounted for. There is no universally agreed definition of carbon footprint yet. Wiedmann and Minx confirmed the same in their study and defined the carbon footprint as a measure of the total amount of carbon dioxide emissions directly and indirectly caused by an activity or accumulated over the life stages of a product.[2]

In terms of carbon footprint, one needs to account for the amount of anthropogenic emissions of GHGs of an activity, business, product or service. The carbon footprint can be measured at a corporate level or for products. The former measurement is termed the corporate or organizational footprint and the latter is termed the product carbon footprint (PCF). Different standards are applied to both kinds of carbon footprints. This entire book is concerned with the concept and measurement of PCF, along with its implications on the textile and clothing supply chain.

As discussed earlier, the carbon footprint is quantified using the GWP indicator over a specific time frame, which is generally 100 years. The quantified result of carbon footprint is expressed in units of mass of carbon dioxide equivalents per unit time or per unit of product (i.e. kg CO_2 equivalents).

3.2.3 Significance of carbon footprint assessment

The reduction of GHGs from various products, processes and organizations has become not only urgent but also increasingly so, as the spike in the concentration of GHGs is affecting our planet in terms of global warming and its associated negative outcomes. The consequences of global warming are numerous and include significant changes in weather patterns, disruption to wildlife, rise in sea levels and undesirable effects on human health.[11] Among these, the impact of climate change on human health is of acute concern and has attracted universal attention.[12–14] The major risks posed by climatic change on human health have both direct and indirect causes and could be[14]

- effects of heat waves and other extreme events (for example, cyclones, floods, storms and wildfires);
- influence on food yields;
- changes in patterns of infectious diseases;
- impact on supply of fresh water and its associated consequences;
- impaired functioning of ecosystems, for instance, wetlands as water filters;
- displacement of vulnerable populations (for example, low-lying island and coastal populations) and a huge loss of livelihoods.[15]

Owing to the damaging effects of global warming, many policies and initiatives have been developed to reduce the emission of GHGs in different countries.

Around the world, governmental, intergovernmental (such as the Kyoto Protocol, IPCC policy) and even various non-governmental initiatives have been set up to support efforts to mitigate global warming. These, along with other measures such as a carbon tax introduced in many countries, encourage producers to take the assessment and mitigation of carbon emissions seriously.

The measurement of the carbon footprint of products, processes and organizations enables an organization to achieve a competitive edge, and those who have measured and are taking steps to reduce their carbon footprint will be set apart from the others. In any industry the company that initiates such standards becomes the leader in their field. Furthermore, although currently carbon footprint value declarations are largely voluntary and driven by self-interest, once they become mandatory, those who have already adopted such standards will be well placed to capture the market. In addition, measuring the carbon footprint provides a way to significantly reduce emissions, as the measurement and management of GHG emissions will undoubtedly help an organization achieve an efficient use of resources, manufacturing processes and reduce waste generation. This in turn leads to improved operational efficiency and cost savings. Measurement of the carbon footprint enables an organization to communicate effectively to its customers about the carbon footprint values of its products, processes and even the organization itself, as well as highlighting its commitment to society in terms of helping achieve a sustainable community.

3.2.4 Product carbon footprint

The basis for measuring a product's carbon footprint lies with the concept of life cycle assessment. In quantifying the carbon footprint of a product, all life cycle phases of a product need to be taken into account, namely, raw material extraction, production of raw materials, production processes of the main product, transportation, distribution and retail, use and finally end of life. The amount of various GHGs emitted will have to be quantified over all these phases and totalled to finally express the particular product's carbon footprint in terms of CO_2 equivalents.

GHG emissions, embodied in all stages from cradle to grave, have to be included for the assessment of the PCF, although sometimes the measurement is confined to the cradle to gate stage alone. Each life cycle stage has many elements and sub-elements linked to it, and all these elements at each life cycle stage of a product have to be included for the assessment of its carbon footprint, depending on the scope and boundary selected for a study (Fig. 3.1). Hence for any PCF study, it is crucial to mention the scope and boundaries of the study, along with other important aspects such as the assumptions made about the functional unit.

3.3 Carbon footprint assessment methodology

In general, whether it is a product or organizational carbon footprint, the rationale behind the measurement is almost identical, consisting of the different steps outlined in Fig. 3.2.

For the organizational footprint, there are three areas that need to be considered in any carbon footprint measurement. The first area deals with direct emissions under the

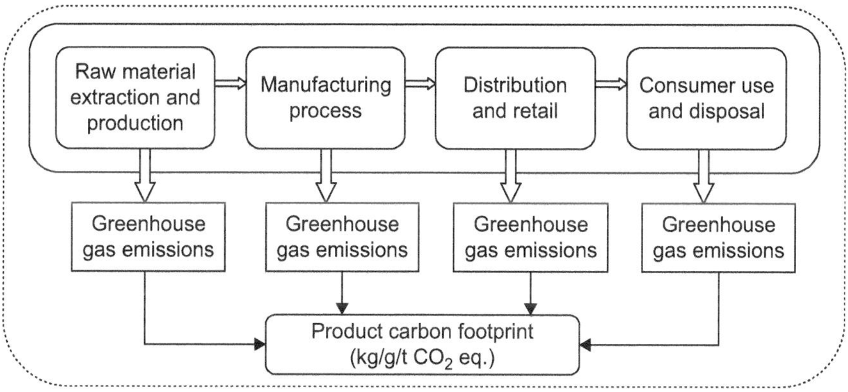

Figure 3.1 The concept of product carbon footprint.

control of the particular organization in which the study takes place (emissions from boilers and furnaces for example), the second area concerns indirect energy emissions (for instance, purchased energy used by the organization) and the third area includes other indirect emissions that do not lie under the control of the organization in question (for example, emissions involved in the extraction of purchased materials). All these need to be defined and clearly identified for an organizational carbon footprint study.

For a PCF assessment, the basic methodology still lies with the concept of life cycle assessment as stated earlier. The study can be confined to cradle to gate stage of a

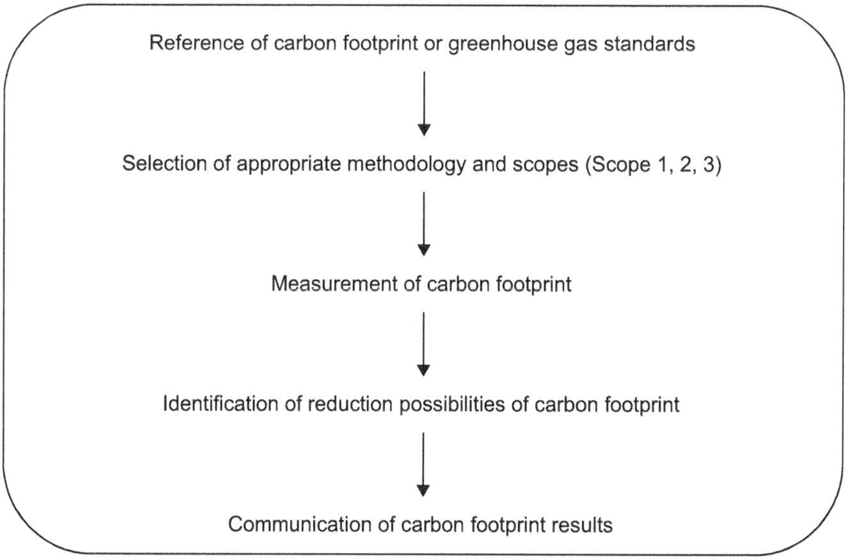

Figure 3.2 Steps involved in assessing carbon footprint.

particular product or it can be extended to its full life cycle from cradle to grave. The key phases of a PCF assessment are shown in terms of a flow chart (Fig. 3.3).[16]

The first step in a PCF study is the selection of appropriate product/s for the study. Numerous factors are considered in this selection, such as whether to choose products

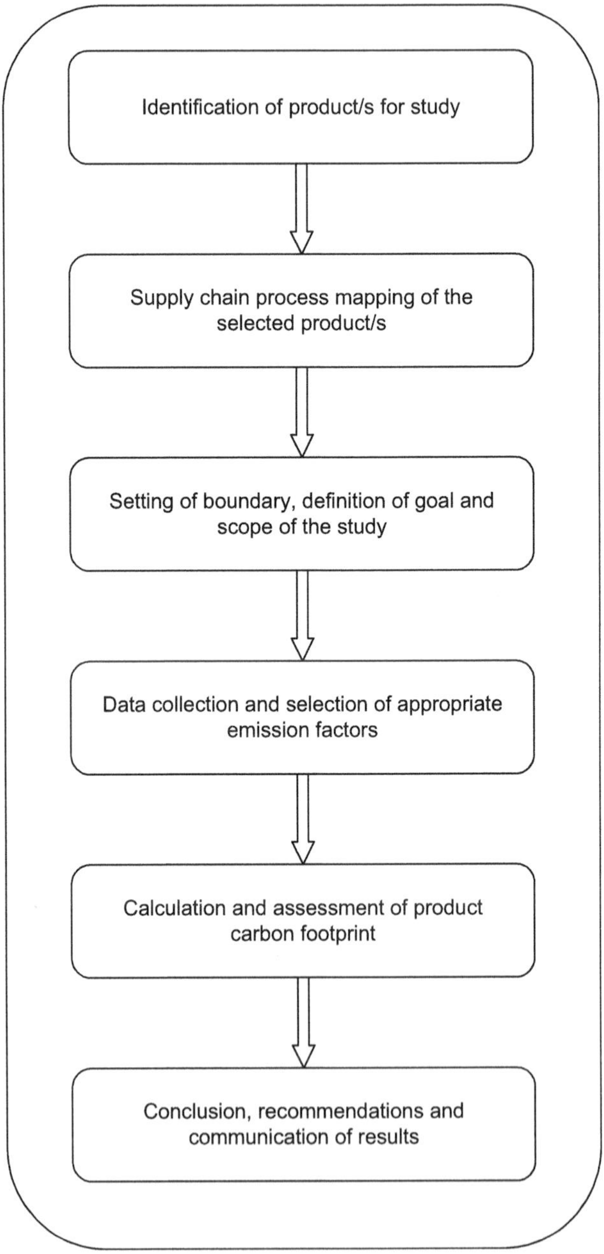

Figure 3.3 Key phases involved in product carbon footprint assessment.

that are perceived to create significant emissions, products that have greater demand in the market or products that customers require a PCF value for.

The next step is to collect relevant product features and raw materials used to manufacture the particular product, sketching the manufacturing flow of the product then mapping the supply chain. Different phases of the life cycle of a particular product selected for this study should be covered with the flow of resources (input and output activities).[16]

The third step is to set the boundary, which means deciding what the study is to cover and what areas of the product's life cycle are excluded. Two important terms widely used in boundary setting are cradle to gate (used in business to business [B2B]) and cradle to grave (used in business to customer [B2C]). B2B type analysis includes all the phases from raw material extraction to manufacturing processes and distribution. Generally, the assessment stops at the point where the product reaches the customer. Sometimes the assessment stops at the factory gate of the organization that is undertaking the PCF study. On the other hand, the B2C study involves the full life of the product from raw material extraction right through the manufacturing process, distribution and retail, consumer use and finally disposal. In this stage, it is also worthwhile defining other aspects such as the goal and scope of the study, the functional unit assumed for the assessment and any other assumptions and limitations of the study.

For any PCF study the next step, i.e. data collection and selection of emission factors, is crucial. The first aspect, activity data, relates to the tangible amounts of material and energy deployed across the various life cycle stages of the product. It also includes the amount and fate of waste produced and transportation details such as the means and distance travelled by both raw materials and auxiliaries, as well as the final product. This can be acquired from primary data sources (i.e. real data collected on-site) or from secondary data sources (i.e. aggregated data collected from secondary sources such as databases and scientific publications). A PCF study must collect the primary data for all possible elements of the life cycle (especially those under the direct control of the organization in which the study takes place, as per publicly available specification [PAS] 2050 standard), and any data gaps due to data that cannot be collected on-site can be filled using secondary data sources.

Emission factors are another aspect of data that play a significant role in the assessment of the PCF. Emission factors provide the basis for the carbon footprint calculation, whether it is for a product or an organization. An emission factor is given as a representative value, which relates the amount or quantity of a pollutant released into the atmosphere with an associated activity for the release of that pollutant. Generally, emission factors are expressed as the mass of a particular pollutant divided by a unit weight, volume, distance or duration of the activity emitting the pollutant (for instance, kilograms of particulate emitted per megagram of coal burned).[17] Emission factors need to be researched and confirmed for every element and activity associated with the different phases of life cycle of a product, based on the selected boundaries of the study.

The fifth step is the assessment of the PCF. In simple terms, assessment involves the calculation of the PCF within the scope and boundaries selected using the collected data. All activity data after conversion to particular units will be multiplied by appropriate and relevant emission factors and added together to produce the final PCF value of the selected product.[16] Once the calculation has been completed, the results of the study will be presented in terms of PCF figures per functional unit of the product assumed for the study. Hence, it is crucial to identify a suitable and appropriate functional unit for any PCF study, as it is defined, following the ISO 14040 standard. The PCF study results need to be presented with sufficient clarity, giving a breakdown of the different elements contributing to the PCF figure, and it is also worthwhile presenting the results in terms of the various life cycle stages selected for the study. This serves as an important aspect of the entire study, as only at this stage can one identify the major contributors to GHG emissions behind the calculated PCF figure, which is vital to further work on their reduction, the main aim of any PCF study. It is also important for a PCF study to deal with scenario setting, studying the effect of different factors on the final PCF result and uncertainty analysis, both of which will be dealt with in subsequent chapters.

A PCF study must also include the observations on-site by the responsible consultant/s or engineers and detailed recommendations to reduce the GHG emissions. Like the goal and scope, the results of the study can be communicated to responsible people such as the management, customers and upstream suppliers. Carbon footprint values have become an important symbol today and their communication to customers is a growing business. Any PCF study should adhere strictly to PCF standards, and any communication should follow the prescribed labelling guidelines to avoid green washing. There are numerous voluntary/mandatory carbon labels available worldwide for the communication of PCF results to the public.

Before communicating the results of the PCF, all the steps detailed earlier have to be followed, but the calculation of PCF is not as simple as might be expected. The calculation involves many steps, including deciding on the methodology to be followed, data collection, selection of appropriate emission factors, allocation of energy and resources to products and co-products, setting of boundaries, decisions as to cut-off rules and calculation of waste.

In order to achieve transparency and uniformity of assessment methodology, several standards have been developed to describe the methodology to be followed for the assessment of PCF. All the PCF standards available are based on both the full life cycle approach and the ISO 14040/44 standards for life cycle assessment. One of the widely used standards is the GHG Protocol, Product Life Cycle Accounting and Reporting Standard published in October 2011. A PAS 2050 is also used in most countries and has been available for a lengthy period. A complete product-based carbon footprint standard ISO 14067 was earmarked and released as a technical specification in 2013 as ISO/TS 14067:2013, which specified principles, requirements and guidelines for the quantification and communication of the carbon footprint of a product, based on the International Standards on life cycle assessment (ISO 14040 and ISO 14044) for quantification and on environmental labels and declarations (ISO 14020, ISO 14024 and ISO 14025) for communication. The ISO/TS 14067:2013 standard was replaced as a complete standard as ISO 14067:2018 in 2018.

3.4 Applications of product carbon footprint assessment to key stages in the supply chain

Carbon footprint (measurement of greenhouse emissions) is one of the essential indicators adopted by major certification schemes and standards pertaining to the textiles industry. This section deals with the assessment and reduction of the PCF in the textiles and clothing supply chain. Every stage in the life cycle of textile products and every element in the entire supply chain of clothing products (from fibre stage until disposal) is responsible for creating GHG emissions. Being a lengthy supply chain, calculating the cradle to grave emissions of textile products can be quite challenging.

To calculate the carbon footprint of the textile supply chain, one has to begin with the raw material extraction stage continuing through all other subsequent stages until disposal. Every textile fibre is unique in terms of its extraction and manufacturing process, and hence the GHGs it emits in its manufacturing stage will be different. After the fibre production process, there are various ways, as described in detail in Chapter 1, to manufacture textile fabrics. Different yarn manufacturing mechanisms and fabric manufacturing technologies will emit varying levels of GHGs. Once the raw fabric is manufactured, the various finishing treatments also have different levels of GHG emissions. The final step of textile product manufacturing, i.e. garmenting or clothing manufacture, is a very similar process (except for variations in the type of sewing machines involved and value-added process additions if any) for all types of textile products. Consequently, GHG emissions of clothing manufacture may not vary so widely for different textile products. After the manufacturing phase, products need to be transported and distributed to the customers. GHG emissions in this phase depend entirely on factors such as distance of transport, mode of transport and the type of fuel being used.

The next phase in the life cycle of textile products is the use phase, which is responsible for creating the majority of GHG emissions. Textile products need to be washed and dried or dry cleaned when used repeatedly, and this maintenance process continues for as long as the functional limit of a textile product permits. GHG emissions during this phase vary very significantly depending on the geographical location in which the product is being used and the associated factors coupled with the maintenance process explained earlier. Consumer behaviour plays a key role in determining the level of GHGs emitted in terms of deciding the lifetime of an article, frequency of washing and drying and selection of washing methods (by machine or hand) and drying techniques. Longevity of a particular product and different technical elements, such as the type of material being used to produce a product and its end-use, are also major factors in terms of the GHG emissions created by a product in the use phase.

After the use phase comes disposal. Various destinations, namely, reusing, recycling, incineration and disposal to landfill, result in varying levels of GHG emissions. The type of material in a disposed product is a common factor influencing the amount of emissions during this phase regardless of the type of disposal. Apart from this, different disposal routes have their own factors affecting GHG emissions.

Along with all the above-mentioned variables, the place in which fibre is produced and where the final textile product is manufactured also play a large part in determining the GHG emissions of a product during its manufacturing phase. Every country has its own specific energy carrier (in terms of available energy source) or energy mix that significantly affects the level of GHGs emitted while manufacturing a product. Different energy sources such as coal, natural gas and oil have different emission levels. Even the same energy source will have differing levels of emissions in different countries, as the quality of energy source available in a particular country and the process of producing energy from a particular energy source, along with various other technical factors, influence the level of emissions. Moreover, each study is unique in terms of its boundary, scope, assumptions and limitations. In view of this, it is not possible to generalize about the GHG emissions of a textile product produced in different countries. Even the results for the same product produced in two adjacent factories in the same city will be different.

3.5 Application of product carbon footprint assessment in textiles

The following sections look at the use of PCF methodology to assess the environmental impact of

- textile products
- raw materials
- manufacturing processes
- distribution operations
- consumer use and disposal.

The discussion starts by looking at the overall product life cycle.

3.5.1 Carbon footprint of a textile product

As stated earlier, the calculation of the carbon footprint of products (textile products here) should follow the life cycle approach, and after completing the calculation, it is possible to determine in which phase of its life cycle the major emissions occur, so analysis can be further extended to discover feasible ways to reduce the same.

A typical textile product starts its life cycle at the fibre production phase, which consists of farming, harvesting and ginning for cotton textiles and in the case of synthetic materials the phase consists of fossil fuel extraction followed by synthetic material processing activities. The fibre production sequence is followed by yarn and fabric manufacturing activities (their process sequences and processing routes depend on the type of product, whether it is woven or knitted and the end-use of the product). After the raw fabric is formed, other processes such as dyeing, finishing and clothing production will be carried out as required to complete the production process. This sequence is explained in detail in Chapter 1.

Once the product is produced and distributed, it reaches the customer and the customer uses the product for as long as the functionality of the product permits.

During this phase, the customer will wash and dry the product to maintain it and at the end of its lifetime will dispose of it as explained in the previous section. The majority of studies investigating the carbon footprint of textile and clothing products concluded that the consumer use phase is the largest contributor of GHG emissions in a typical product's life cycle, and some of the studies also found that fabric production occupies a dominant part in the total life cycle GHG emissions of a product. Aggregated GHG emissions of a typical textile product across different life cycle phases are presented in Table 3.1.

Although Table 3.1 cannot represent the exact situation of all textile products (and the GHG emissions across different life cycle phases vary for different products based on numerous factors, as already discussed), it gives an overall picture of a textile product in terms of GHG emissions during its life cycle.[18] It is clear from Table 3.1 and many other studies conducted to date that the majority of GHG emissions for clothing products lie with customers in their use phase. Clothing products require washing, drying and ironing, needing higher energy for these use phase activities and consequently the use phase alone contributes to around 40%–80% of the total GHG emissions arising from such a clothing product.[18] Needless to say, garments requiring only handwashing will generate fewer GHG emissions in the use phase. Some of the studies also found other life cycle phases, such as fabric production, to be a major hot spot in the total life cycle emissions of a textile product. This will be highlighted in one of the later chapters that deals with case studies of textile products. For instance, one of the studies reported in 2012 based on UK clothing found that the fabric production that included weaving/knitting and treatment of fabric contributed to 33% of total life cycle GHG impacts, followed by the use phase that was found to be responsible for 26% of total

Table 3.1 Total aggregated GHG emissions of multiple clothing types.[18]

Life cycle phase	% Total GHG emissions (for multiple clothing types)
Fibre production	18
Yarn production	16
Preparation and blending	5
Knitting	7
Dyeing and finishing	3
Other raw materials	5
Garment manufacture (making up)	2
Packaging	4
Transportation	1
Use phase	39

GHG, greenhouse gas.

life cycle GHG impacts. Furthermore, out of the activities in the use phase, washing was found to be the largest contributor (15%), followed by drying (10%) and ironing (1%).[19]

3.5.2 Carbon footprint of textile raw materials

A wide array of raw materials are used to manufacture textile products in the clothing sector. Textile raw materials (of both natural and synthetic origin) include, but are not limited to, conventional cotton, organic cotton, flax, acrylic, viscose, polyester, polypropylene, nylon, silk and wool. Their manufacturing processes and the implications of the same on the environment were briefly discussed in Chapter 1. Fibres of either natural or synthetic origin require enormous levels of energy and are manufactured by different processes, producing varying levels of GHG emissions. Even the same type of fibre produced in different countries will have different carbon footprints because of the reasons already discussed in this chapter. For example, organic cotton produced in the United States will generate a different carbon footprint to one produced in India or China. Again, the type of fibre used to produce a particular garment will affect the GHG emissions in subsequent life cycle stages as well, for instance, emissions during the manufacturing and use phases of a cotton T-shirt will be different from those of a polyester T-shirt.

Different studies have identified the GHG emissions of various fibres from cradle to gate stages and even cradle to grave stages. Muthu *et al.*[20] used GHG emissions from cradle to gate stages for different fibres in their unique model developed to quantify the environmental impact and ecological sustainability indices of textile raw materials. GHG emissions of textile fibres from various studies used by Muthu and co-workers in their study are given in Table 3.2.

A study conducted in the United Kingdom in 2009 calculated the total quantities of all major types of fibre (cotton, acrylic, linen, nylon, polyester, polypropylene, silk, viscose and wool) purchased and consumed in the United Kingdom. This study accumulated data from various sources for all these fibres and the different stages of manufacturing processes of garments made from these fibres.[19] Table 3.3 extracted from this study lists the carbon footprint results of the different textile fibres included in the study.

The results in Table 3.3 show that wool was responsible for the highest carbon emissions, due to methane emissions from sheep and manure, along with nitrous oxide emissions from fertilizers.[18,19] Flax recorded a very small carbon footprint because linen was allocated as a low-value co-product from linseed oil.[19] However, even apart from this study, linen generally has lower GHG emissions owing to low irrigation and pesticide requirements.[18]

3.5.3 Carbon footprint of textile manufacturing processes

There are many processes in the clothing supply chain to convert the raw material to a finished product. The different process routes for various fibre types and for different applications were dealt with in Chapter 1. In general the main processes include yarn

Table 3.2 Greenhouse gas emissions of various textile fibres.[20]

Fibre	CO_2 emissions (kg CO_2e per kg of fibre)
Nylon 6	5.5[21]
Nylon 66	6.5[21]
Viscose	9 (−3.5 for bio-mass credit)[22]
Acrylic	5[22]
Polyester	2.8[21]
Organic cotton	2.5[22]
Conventional cotton	6[22]
Wool	2.2[22]
Flax	3.8[23]
Polypropylene	1.7[21]

spinning, weaving or knitting; wet processes such as singeing, bleaching, dyeing, printing, finishing (mechanical and chemical); and finally clothing manufacturing. Most of these processes are energy intensive and hence are responsible for a significant amount of GHG emissions. Apart from their energy needs, these processes also require a lot of consumables that indirectly contribute to the emission of GHGs.

In one of the studies conducted on a long white shirt, it was discovered that it emitted 10.75 kg CO_2e (weighing 222 g and made from 100% cotton, its carbon footprint was 50 times its weight). Of those emissions, 3 kg CO_2e (around 28%) arose from the manufacturing processes. Further analysis of the manufacturing process emissions

Table 3.3 Greenhouse gas emissions of various textile fibres from a UK study.[19]

Fibre	CO_2 emissions (kg CO_2e per tonne of fibre)
Nylon	8070
Viscose	2118
Acrylic	7577
Polyester	5357
Silk	2031
Cotton	1755
Wool	20,790
Flax	335
Polypropylene	3097

revealed that the spinning process emitted 1.02 kg CO_2e, knitting emitted 0.12 kg CO_2e, while dyeing and garmenting processes were responsible for 0.98 kg CO_2e and 0.88 kg CO_2e, respectively.[23]

For this particular garment, spinning was the primary hot spot contributing 34%, followed by the dyeing and garmenting processes with 33% and around 30%, respectively. Thus there was no significant difference between the contributions of these processes. For this product, the lowest contribution came from the knitting process, with just 4%. It is understandable that knitting produces a lower level of emissions because of the very limited amount of energy and resources involved (yarn is immediately converted to fabric in knitting by using only one machine without any processes in-between). Consequently, the GHG emissions from the knitting process should be low. Apart from the technical boundaries discussed earlier, the country in which the product is made affects its carbon footprint. Factors such as the method of production of electricity in a particular factory in that country and the availability of energy sources to produce electricity in a specific country alter the emissions arising during the manufacturing processes.

This particular case study was conducted in Bangladesh, which had a deficiency in its grid supply and hence the suppliers had to rely on self-production on-site for most of their energy requirements. But again, natural gas is abundantly available in Bangladesh and generates lower GHG emissions than coal or oil,[24] so, as previously mentioned, generalizations cannot be made for all textile and clothing products.

Another case study on a T-shirt manufactured in South India found that the manufacturing process alone contributed to around 19,080.4 kg CO_2e of the total emissions of 21,922.80 kg CO_2e per tonne of T-shirt.[25] On a cradle to gate analysis of the T-shirt, it was established that 87% of emissions were produced during manufacturing processes, and of the manufacturing emissions, spinning was responsible for 2787 kg CO_2e (around 15%) and knitting 253.8 kg CO_2e (1.3%). The dyeing process was the major contributor to this product's GHG emissions, producing 15,734.8 kg CO_2e (around 83%), and the remaining 304.8 kg CO_2e (around 1.6%) arose during the garmenting process. This makes clear that dyeing is the major hot spot in the manufacturing process for a T-shirt, followed by spinning. It also confirms that the knitting process is responsible for fewer carbon emissions. The weaving process will cause higher levels of carbon emissions than knitting because of the many processes involved between yarn and fabric formation, such as winding, warping, sizing and finally weaving. These processes demand a higher energy profile and require many consumables, consequently their carbon emissions will be higher.

3.5.4 Carbon footprint of transportation and distribution phases

The impact of transportation and distribution depends on many factors (as discussed already) such as the distance and mode of transport. For the example of the long white shirt, transportation was responsible for 0.29 kg CO_2e (3%).[25] This was for sea freight transport from Asia to Europe; however, if it had been transported by air then the impact on transportation would have gone up to 4 kg CO_2e.[26] There is clearly a noticeable difference between 0.29 and 4 kg CO_2e between modes of transport over the same distance. For the same product the distribution process produced 0.87 kg CO_2e (8%) of emissions, which is even higher than the transportation emissions.

3.5.5 Carbon footprint of consumer use and disposal phases

Consumer use is one of the most important phases of a textile product's life cycle, where the majority of carbon emissions occur. Hence the consumer's attitude and behaviour towards the care and maintenance of textile products could reduce the carbon emissions during the use phase. With the exception of single-use textile products (such as nappies) or products that do not require maintenance and care, all other textile products will have significant emissions arising over the use phase.

In the case study of the long white shirt, it was revealed that one-third of carbon emissions of the product's total life cycle lay in the consumer use phase, assuming 55 laundry cycles over its entire life time. The use phase alone contributed 3.3 kg CO_2e (31%) of emissions in this product. Carbon emissions of the use phase depend mainly on washing and drying frequency over the lifetime of a garment, the chosen temperature for washing, energy efficiency of the washing machine and dryer used and the degree of load during washing and drying. From this study, it was calculated on a scenario analysis that if a dryer had been used for each laundry cycle then the carbon emissions of the use phase alone would have increased to 10 kg CO_2e from 3.3 kg CO_2e.[24] Not only this study but also many others have dealt with different scenario settings to emphasize the importance of this phase in reducing carbon emissions of textile products.

Different types of garments made from various fibres have specific washing and drying requirements, as every fibre behaves differently during washing and drying. A UK study conducted in 2009, which was discussed earlier, calculated the carbon footprint of ten types of textile products made from different types of fibres, namely, cotton, acrylic, linen, nylon, polyester, polypropylene, silk, viscose and wool. This was a comprehensive study dealing with many aspects of carbon footprint analysis of these ten selected garments from different angles and under various scenarios. One of the analyses quantified the carbon footprint of each of these garments at various life cycle phases from cradle to grave. The results of the carbon footprint of these garments over the use and disposal phases are listed in Table 3.4.

Table 3.4 clearly demonstrates that different types of garments made from various fibre types emit different levels of GHGs due to the reasons already discussed. Similar to the use phase, the disposal phase will vary in terms of the emissions arising from different garments. As discussed earlier, once a customer decides to dispose of a textile product, it may go to one of the several destinations, namely, reuse, recycling, incineration or disposal to landfill. Again, recycling has two variants, namely, open and closed loop recycling, and both of them have varying levels of carbon emissions. The level of carbon emissions for a textile product will depend on the chosen destination of disposal. In many cases, instead of only one destination, an end-of-life scenario comprises mixed destinations. Indeed in this UK study, an end-of-life scenario

Table 3.4 Greenhouse gas emissions of ten garments from a UK study.[19]

	CO_2 emissions (kg CO_2e per garment)			
Garment type	**Washing**	**Drying**	**Ironing**	**Disposal**
Tops	0.9	0.6	0.0	−0.5
Underwear, nightwear	0.3	0.2	0.0	−0.1
Bottoms	1.3	0.9	0.1	−0.6
Jacket	1.9	1.3	0.1	−1.0
Dresses	2.6	1.7	0.1	−1.0
Suits	2.1	1.4	0.1	−1.3
Gloves	0.1	0.1	0.0	−0.1
Sportswear	1.1	0.7	0.0	−0.4
Swimwear	0.3	0.2	0.0	−0.1
Scarves, shawls and ties	0.2	0.2	0.0	−0.1

comprising different destinations was chosen. The implications of different disposal routes on the environment will be discussed in detail in Chapter 8.

A study, Environmental Impact of the Global Apparel and Footwear Industries Study, which was published in 2018, conducted a life cycle assessment and measured five impact indicators and one of them is the carbon footprint. Functional unit assumed in this study was the global apparel annual production and the chosen impacts were likewise calculated based on the global apparel yearly consumption per capita. This study did not include use phase, packaging, cleaning and ironing and accessories. According to the results of this study, total climate change impacts expressed in gigatons CO_2 equivalents are 3.29 and out of which fibre production was responsible for 0.51; yarn production, 0.93; fabric production, 0.39; dyeing and finishing, 1.18; assembly stage, 0.22; and distribution, 0.04. Dyeing and finishing (responsible for 36%) was found to be the major hot spot in terms of carbon footprint, followed by yarn production (responsible for 28%).[26]

3.6 Summary: key challenges in calculating the product carbon footprint in textiles

Environmental regulations in different countries across the globe require organizations to measure and declare their carbon footprint at organizational and product levels. Customer demand also drives the assessment of PCF and the declaration of values in terms of PCF labels. Apart from the pressures from competition, regulation and customer demand, an organization needs carbon footprint measurements to reduce carbon emissions and costs by improving operational efficiency, and thus contribute towards achieving a sustainable society.

There are ample opportunities for the assessment and declaration of carbon footprint results. By assessing its carbon footprint, an organization meets regulatory guidelines (whether existing or potential) and customer demands and demonstrates a desire to curb carbon emissions and take the lead in a low-carbon world. Apart from these opportunities, a company can identify problematic areas that are responsible for higher GHG emissions and work to reduce them.

In addition to opportunities, challenges exist in assessing the PCF of textiles and clothing products. The challenges of a PCF study are shown in the following range:

- choosing a suitable methodology/standard for the study;
- selecting appropriate emission factors;
- dealing with the choices of data and data quality;
- dealing with data uncertainty;
- technical know-how in data collection, filling of data gaps with appropriate data;
- selection and availability of appropriate secondary data, if primary data is not available;
- tackling methodological complexities, such as allocation of products and co-products, dealing with cut-off rules of the study;
- mapping of boundaries and inclusion and exclusion of different areas of consideration, especially for a longer supply chain such as textiles;
- selecting one or more products from those available on the shelf;
- as the PCF result of one product is unique to that product, difficulties in assessing a large number of products (where there are many changes in design and product profiles) vary frequently, such as for textiles and clothing products;
- availability of suitable databases for textiles and clothing products and their life cycle phases;
- verifying and validating results, if not done by an approved third party;
- a typical issue, related to the textile sector, is that calculated products keep changing with various product profiles requiring a reassessment of the PCF calculation.

This chapter dealt with a comprehensive overview of the main principles underlying the measurement of the carbon footprint, emphasizing the scientific concepts of GWP, and gave a brief introduction to the emerging concept of carbon footprint measurement. It also enumerated the significance of carbon footprint measurement and the communication of carbon footprint results of products and organizations to different interested parties. Having discussed the details of organizational and PCFs and their measurement principles, it dealt with the step by step procedure of conducting a PCF study, along with the various standards applicable for the measurement of a PCF.

Following this methodological introduction of GHG measurement and its underlying principles, the chapter dealt with the application of carbon footprint measurement to the key stages of the textile and clothing supply chain. An overview of a textile product's carbon emissions over its entire life cycle from cradle to grave stages was presented. Also, carbon emissions at important life cycle phases of textile and clothing products (namely, the raw materials stage, manufacturing processes, transportation and distribution phases, consumer use and disposal phases) were discussed with relevant case studies. Finally, an important part of this chapter highlighted the key challenges faced by researchers and engineers who attempt to measure the carbon footprint of textile and clothing products.

3.7 Sources of further information and advice

- Barber A, Pellow G. *Life Cycle Assessment: New Zealand Merino Industry, Merino Wool Total Energy Use, Carbon-Dioxide Emissions*. Auckland: The Agri Business Group; 2006.
- Baumert KA, Herzog T, Pershing J. *Navigating the Numbers: Greenhouse Gas Data and Climate Change Policy*. Washington, DC: World Resources Institute; 2005.
- Fletcher K. *Sustainable Fashion and Textiles*. London: Earthscan; 2008.
- Herrmann IT, Hauschild MZ. Effects of globalisation on carbon footprints of products. *CIRP Annals—Manufacturing Technology*. 2009;58:13–16.
- Kerr AR. A worrying trend of less ice, higher seas. *Science*. 2006;311:1698–1701.
- Lash J, Wellington F. Competitive advantage on a warming planet. *Harvard Business Review*. 2007;85(3):94–102.
- Page EA. Distributing the burdens of climate change. *Environmental Politics*. 2008;17(4): 556–575.
- Sundarakani B, Souza RD, Goh M, Wagner SM, Manikandan S. Modeling carbon footprints across the supply chain. *International Journal of Production Economics*. 2010;1–4.
- Weidema BP, Thrane M, Christensen P, Schmidt J, Løkke S. Carbon footprint: a catalyst for life cycle assessment. *Journal of Industrial Ecology*. 2008;12(1):3–6.
- Wiedmann T, Minx J. *A Definition of Carbon Footprint*. ISAUK Research Report 07–01. ISAUK Research and Consulting: Durham; 2007.
- Wright LA, Kemp S, Williams I. 'Carbon footprinting': towards a universally accepted definition. *Carbon Management*. 2011;2(1):61–72.

3.8 References

1. *Climate Change Basics*. Available at: http://www.epa.gov/climatechange/basics/.
2. IPCC. *Climate Change 2007: Synthesis Report: Contribution of Working Groups I, II and III to the Fourth Assessment Report*. Intergovernmental Panel on Climate Change; 2007.
3. Kerr AR. How urgent is climate change? *Science*. 2007;318:1230–1231.
4. United Nations. *Framework Convention on Climate Change*. Available at: http://unfccc.int/kyoto_protocol/items/3145.php.
5. *Information on Kyoto Protocol*. Available at: http://webarchive.nationalarchives.gov.uk/+http://www.bis.gov.uk/policies/business-sectors/ccpo/kyoto-protocol.
6. *Climate Change 2007: Working Group I: The Physical Science Basis*. Available at: http://www.ipcc.ch/publications_and_data/ar4/wg1/en/ch2s2-10.html.
7. *Global Warming Potential*. Stockholm Environment Institute, Carbon Offset Research & Education (CORE), GHG Management Institute. Available at: http://www.co2offsetresearch.org/aviation/GWP.html.
8. Forster P, Ramaswamy V, Artaxo P, Berntsen T, Betts R, et al. Changes in atmospheric constituents and in radiative forcing. In: Solomon S, Qin D, Manning M, et al., eds. *Climate Change 2007: The Physical Science Basis. Contribution of Working Group I to the Fourth Assessment Report of the Intergovernmental Panel on Climate Change*. Cambridge, UK and New York: Cambridge University Press; 2007.
9. Rees WE. Ecological footprints and appropriated carrying capacity: what urban economics leaves out. *Environment and Urbanization*. 1992;4(2):121–130.
10. Wackernagel M, Rees WE. *Our Ecological Footprint: Reducing Human Impact on the Earth*. Gabriola Island: New Society Publishers; 1996.

11. *Positives and Negatives of Global Warming*. Available at: http://www.skepticalscience.com/global-warming-positives-negatives.htm.
12. McMichael AJ, Woodruff RE, Hales S. Climate change and human health: present and future risks. *Lancet*. 2006;367:859−869.
13. Haines A, Kovats RS, Lendrum CD, Corvalan C. Climate change and human health: impacts, vulnerability, and mitigation. *Lancet*. 2006;36:2101−2109.
14. Subramanian SM, Li Y, Hu JY, Mok PY. Carbon footprint of shopping (grocery) bags in China, Hong Kong and India. *Atmospheric Environment*. 2011;45:469−475.
15. McMichael AJ, Friel S, Nyong A, Corvalan C. Global environmental change and health: impacts, inequalities, and the health sector. *BMJ*. 2008;336:191−194. https://doi.org/10.1136/bmj.39392.473727.AD.
16. Rugrungruang F, Chua BH, Low SC. Development of product carbon footprint assessment: a step towards sustainability for Singapore manufacturing industry. SIMTech Technical Reports *(STR_V10_N2_09_STA)*. April−June 2009;10(2).
17. Information on emission factors, referred from the page of *Emissions Factors & AP 42, Compilation of Air Pollutant Emission Factors*. Available at: http://www.epa.gov/ttnchie1/ap42/.
18. *Apparel Industry Life Cycle Carbon Mapping, Prepared by Business for Social Responsibility*. Business for Social Responsibility (BSR); June 2009. Available at: http://www.bsr.org/reports/BSR_Apparel_Supply_Chain_Carbon_Report.pdf.
19. Thomas B, Fishwick M, Joyce J, van Santen A. *A Carbon Footprint for UK Clothing and Opportunities for Savings*. WRAP, Final Report; July 2012. Available at: http://www.wrap.org.uk/sites/files/wrap/Appendix%20IV%20-%20Carbon%20footprint%20report.pdf.
20. Muthu SS, Li Y, Hu JY, Mok PY. Quantification of environmental impact and ecological sustainability for textile fibres. *Ecological Indicators*. 2012;13(1):66−74.
21. Morris D. The fibres, textile and textile manufacturing industries in China P.R. forecasts and environmental considerations. In: *77th International Wool Textile Organisation Congress, Beijing, China*. Available at: http://www.cirfs.org/press/TheFibres,TextilesandTextileManufacturingIndustries.pdf.
22. Cherrett N, Barrett J, Clemett A, Chadwick M, Chadwick MJ. *Ecological Footprint and Water Analysis of Cotton Hemp and Polyester*. Stockholm, Sweden: Stockholm Environment Institute; 2005.
23. Jungmichel N. *The Carbon Footprint of Textiles*. Berlin: Systain Consulting; 2010. Available at: http://www.ci-romero.de/fileadmin/media/informieren-themen/gruene_mode/Jungmichel.Systain.pdf.
24. *Carbon Footprint Study 2009 Final Summary, 220 grams Textile, 11 kilograms CO2, The Carbon Footprint of Clothing*. Systain Consulting. Available at: http://www.systain.com/fileadmin/Dateien_Systain/Daten/Download_Dokumente/sys_Studie_carbonfootprint_english.pdf.
25. Nagel S, Schoeman C, Yadavalli V. Exploring ways to reduce the greenhouse gas emissions in the textile supply chain. In: *SAIIE Annual Conference, North America, September 2010*; 2010. Available at: http://www.saiie.co.za/ocs/index.php/saiie/SAIIE10/paper/view/161. Accessed March 21, 2013.
26. Measuring Fashion. *Environmental Impact of the Global Apparel and Footwear Industries Study, Full Report and Methodological Considerations*. Quantis; 2018.

Calculating the water and energy footprints of textile products

4

4.1 Introduction

After measuring carbon footprints, measurement of water and energy footprints is crucial for industries, including the textiles and clothing sector. Water and energy are two major resources that are depleting rapidly but are consumed by almost all industrial sectors. Water is one of the basic necessities of human life and its scarcity in the future is emerging as an issue for many countries. Energy is another vital resource to be preserved for our future generations to achieve a sustainable society.

The textile supply chain needs a huge amount of water and energy to manufacture a textile product. Apart from production, a typical textile product needs large amounts of both water and energy in its use phase too. Hence, determination of water and energy footprints in a textile product's life cycle is vital and seeking ways to reduce consumption of both water and energy are highly relevant. This chapter deals with the basic concepts and measurement techniques of water and energy footprints and their application to the textile and clothing supply chain. It also highlights the main challenges in measuring water and energy footprints of textile products.

4.2 Water footprints: an introduction

The importance of fresh water is self-evident, and it is well known that owing to the spike in population growth, all essential resources, including fresh water, are becoming scarce. Water scarcity is the failure to meet the water demand by freshwater resources. In the recent days, especially now, water scarcity is one of the main issues in many countries in South East Asia, including India, which is one of the major textile production hubs in South East Asia. Recently, say around within the past 100 years, water use is growing double the rate than the growth of population and hence it is becoming one of the top most global issues. According to the very recent data from the World Resources Institute (WRI), as of 2019, 17 countries in total are experiencing 'extremely high' levels of baseline water stress and 44 countries are experiencing 'high' levels of water stress.[1]

Fresh water serves many purposes, such as in agricultural use, production of hydroelectric power, industrial processes and provides drinking water. It is a vital resource for the survival of human beings and is required in almost all industrial processes. Again, looking simply at the product level, any product needs fresh water in its complete life cycle from cradle to grave. This is certain but the means to satisfy it are not. The primary issue is the imbalance of demand caused by rising levels of population and the availability of fresh water.

Assessing the Environmental Impact of Textiles and the Clothing Supply Chain
https://doi.org/10.1016/B978-0-12-819783-7.00004-1

According to the statistics from the World Health Organization, one-fifth of the entire population in the world, which accounts to more than 1.1 billion people, live in water-scarce areas. It is projected by the UN that by 2025, around 1.8 billion people will suffer from water scarcity.[2] As said earlier, with the available resources, the fresh-water needs of the entire human population cannot be met. In continuation to the recent data explained earlier, nearly one-quarter of the world's population, which comes to say around 1.7 billion people, currently live in areas where agriculture, industries and cities withdraw 80% of their available water supply every year.[1] In 1995, 3906 km^3 of water was withdrawn globally to meet all the demands listed earlier, such as drinking, food security, agricultural processes and industrial processes. It is projected that by 2025 the need for water for these various uses will increase by at least 50%.[3] Hence, it is clearly important to measure and reduce water consumption.

This brief introductory outlook highlights the need for a water footprint (WF). The concept of a WF, which helps illustrate the consumption of water resources on a global level,[4] was first introduced in 2002 by Hoekstra and Hung,[4] and further discussed by Chapagain and Hoekstra in 2004.[5] It was developed analogous to the ecological footprint, which had been introduced in the 1990s. The WF quantifies direct and indirect water usage by consumption or manufacture of products and or services.[6–8] It can be measured for a business, a community or an individual, being the total amount of fresh water used to produce the products or goods and services consumed by the community or individual or produced by the business activity. Water usage is measured in terms of the volume of water evaporated or incorporated into a product and/or polluted per unit time.[9]

The idea of a WF is very closely related to the virtual water theory. Virtual water is defined as the amount of water embedded in food or the volume or amount of water utilized to produce a particular product.[8,10] The concept of virtual water was introduced in the 1990s by Allan to solve water scarcity issues by importing virtual water as opposed to real water.[11]

A WF can be divided into three elements:

- Blue WF—refers to the volume of surface water and groundwater consumed during production processes.
- Green WF—refers to the volume of rain water consumed.
- Grey WF—refers to the quantity of fresh water needed to mix desirable pollutants and maintain the required water quality as prescribed by agreed water quality standards.[12]

The WF can be measured and expressed per day, month or year depending on the level of information needed,[13] and the WF of products is always expressed as water volume per product unit. Some of the examples are water volume per unit of mass (litre/kg or m^3/t of products) if weight is chosen as a quantity indicator; then the water volume per piece of the product or number of pieces is measured. If the WF of processes is measured, it is usually expressed as water volume per unit of time. When this unit is divided by the units of products measured during that time, then it can also be expressed as water volume per unit of product.[8]

4.3 Methods for assessing water footprints

As mentioned earlier, the WF of a product quantifies the amount of fresh water used to produce a product over the entire supply chain. Assessment of the WF refers to the measurement and location of the WF of a particular process, product, producer and/or consumer and measures its sustainability (covering all the three pillars of sustainability). A response strategy is then followed.[14] Assessment of the WF aids in our understanding of the consumption of fresh water by specific human activities or the production processes of a product and how they contribute to issues of water scarcity and pollution. It also helps formulate a sustainable policy to continue those human activities or production processes.

WF can be measured for a product by covering all aspects of its supply chain or a specific activity or process of a lengthy supply chain. Similar to life cycle assessment (LCA), WF assessment consists of four phases:

1. setting goals and scope (Phase 1),
2. WF accounting (Phase 2),
3. WF sustainability assessment (Phase 3),
4. WF response formulation (Phase 4).[13]

The four phases in WF assessment are similar to those of an LCA study as per ISO 14040/44 standards. Setting goals and scope is the first step. Every study has a single objective but there may be many goals such as the study of the WF of a complete product or one particular process of the supply chain or of a consumer or a group of consumers in a particular nation or locality. WF can also be measured for a business or an entire business sector. The main aim of a WF study, whether raising awareness amongst the public, hot spot identification, setting targets or aiding policy formulation, has to be clearly stated. Also, any specific focus of interest of the study, such as direct/indirect footprints and grey/blue and/or green footprints, must be indicated, and other details such as the time frame within which the study is to be completed must also be specified.[14]

The second phase is WF accounting. A WF study should have clearly defined boundaries with clearly identified areas of inclusion and exclusion. Different considerations in terms of the following checklist must be covered and indicated in the study too.

- Consideration of different components of a WF—blue/grey and/or green footprints?
- Which period of data should be included?
- Consideration of only direct/indirect footprints or both in the study?
- Point of cut-off when considering the entire supply chain?
- For a national WF, whether only part of a nation or entire national consumption is considered?

The third phase, and a vital one in any study, is the WF sustainability assessment. The main point to be discussed here depends on the perspective of the WF study; whether from a geographical or a product or process perspective. Sustainability of the aggregated WF in a certain locality needs to be looked at in the case of a geographic perspective WF, while contribution to the WF by the individual product or process has to be the focus in the case of a product or process perspective WF study.[14] This assessment in the third phase of any study is closely linked to the first step of goal and scope definition. Key considerations include whether all facets of sustainability (economic, social and environmental) need to be looked at blue/grey and or green footprints and to simply identify hot spots or further define whether those hot spots are primary or secondary.

While studying the WF of a specific product or process, a study should look at two important aspects:

1. Whether the WF of a product or process gratuitously contributes to the global WF of humanity, by comparing the global average (or benchmark) to that product or process. If such a benchmark is not available, the scope of any study should be extended; therefore, this study also includes understanding what a reasonable benchmark could be.
2. Whether the WF contributes to any specific hot spots by checking whether each component WF lies in a hot spot or not. This requires a worldwide hot-spot database at spatial and temporal levels. If such a database is not available, the scope of any study must be extended to include catchment studies from the geographical perspective as well.[14]

The final phase is WF response formulation, which is also linked to the initial phase of goal and scope definition. In this phase, strategies or options to reduce WF will be formulated. Depending on the scope of any specific study, responsible persons (whether on a national or an individual level) should be alerted as to what should be done to reduce WF.[14]

Detailed steps for calculating the WF of various entities are discussed fully in the WF assessment manual[13] and readers are advised to refer the manual for more detail. The WF of a product can be measured in one of the two ways, namely, chain summation and the stepwise accumulation approach. The former method is very simple, but it can only be applied to a system that produces only one output product. With this method the WF of the product in question is equal to the sum of the WFs of the relevant production processes to produce it divided by the production quantity of that particular product. The stepwise accumulation approach calculates the WF of the final output product as the sum of all the WFs of all input products and the WF of the relevant processes. This is applicable when many input products are required to produce one output product. On the other hand, if one input product is used to produce many output products, the WF of the input product has to be attributed to its separate products (proportionally to the value of the various output products).

The WF of a product can also be calculated using the LCA approach by mapping the entire product life cycle from cradle to grave, accounting for the water usage at different life cycle stages. This method deals with the quantification of fresh water needed to produce a product, including water used in the manufacturing processes, in background processes such as extraction of raw materials, in electricity generation and during the other life cycle stages (use and disposal). Many other methods have been developed modelled on the LCA technique addressing water use.[15] There is a unique standard, ISO 14046 (ISO 14046:2014 Environmental management – Water footprint – Principles, requirements and guidelines) to calculate the water use of products, processes and organizations. This standard is based on ISO 14040, but it focuses chiefly on water use.

4.4 Applications of water footprint assessment to key stages in the textile supply chain

The entire supply chain of textiles and clothing production requires enormous amounts of water at various stages. As with other indicators, it is impossible to generalize about the WF. Factors such as the geographical location in which the product (and other associated materials) is made and the availability and source of water in that location vary. Results will also vary in different studies, as every study is unique in terms of its assumptions and boundaries.

The global average WF of a cotton fabric is 10,000 litres per 1 kg of cotton fabric.[16] However, depending on the geographical location, WF results vary significantly. Based on the studies by Mekonnen and Hoekstra[17,18] in 2010 and 2011, the WF of 1 kg of cotton fabric made from Chinese, American, Indian, Pakistani and Uzbekistan cottons is 6000, 8100, 22,500, 9600 and 9200 litres, respectively.

The production of cotton contributes very significantly to the blue WF since it is frequently irrigated. Blue water makes up almost one-third of cotton's WF. Again, the quantity of blue water varies geographically. In some places, its proportion is much larger, such as in Pakistan and Uzbekistan where blue water contributes 55% and 88%, respectively, to the total WF of cotton production. During the period 1996–2005, the world cotton production alone contributed to 3% of the total WF of global crop production.[17,18]

Cotton is an important natural fibre composing a significant proportion of the textile supply chain and around a 36% share in the entire textile fibre market.[19] Production of cotton fabrics not only causes depletion of resources but also pollutes resources at both agricultural (field level production of cotton) and industrial levels (processing of seed cotton for further conversion into final cotton products). A research study by Chapagain and colleagues in 2006 dealt with the WF of worldwide cotton production and consumption revealing that within the period covered by this study (1997–2001), worldwide consumption of cotton products required 256 g^3 of water per year. Out of this 256 g^3, 42% was blue water (soil water taken up by plants from infiltrated irrigation water), 39% was green water (soil water taken up by plants from infiltrated rain water) and the remaining 19% was dilution water (water required to assimilate the pollution).[4]

Various measures pertaining to water use in cotton production were considered in this study. Consumptive water use at field level for cotton production, along with the volume of water used and virtual water content of seed cotton in various cotton-producing countries, was studied. The virtual water content of different cotton products at various production stages was also recorded individually for different countries, along with the worldwide figure. The global average virtual water content of cotton lint, grey fabric, finished fabric and final textile products was each presented in terms of blue and green WFs. It was found that the global average virtual water content of a final textile product was 9359 m^3/ton, and on average, $30,177 \cdot 10^6$ m^3 per year of dilution water was required globally to dilute the fertilizers leached to the water bodies.[4] Another interesting aspect reported in this work is the global average virtual water content of some textile products. One pair of jeans weighing 1 kg has 10,850 litres of virtual water content, one T-shirt weighing 250 g has 2720 litres and one nappy weighing 75 g has 810 litres. The total worldwide WF related to consumption of cotton products is 256,336 mm^3 per year.

It is apparent that apart from the application of fertilizers, wet processing is one of the major consumers of dilution water. It was found that the average volume of water used in wet processing was 496 m^3/t (bleaching, dyeing and printing needed 360 m^3/t and finishing needed 136 m^3/t).[4] However, the study of Chapagain and co-workers[4] in 2006 found that 640 m^3/t of dilution water was required to dilute pollution for wet processing and 240 m^3/t of dilution water was required for finishing alone.

Another study conducted by the Stockholm Environmental Institute recorded the water requirements for cotton, hemp and polyester. According to this study, around 9958 litres of water is required to produce 1 kg of cotton and 2123 litres of water is required to produce 1 kg of hemp.[20] It was noticed that polyester requires less than 0.1% of the water required for cotton[21] and 1 kg of polyester production needed only 62 litres of water.[22]

Water use in terms of the entire life cycle stages of a textile product is very important and has to be discussed. Levi Strauss and Company conducted an LCA study in 2007 recording the water use in the production of a pair of their jeans. The study found that over 3000 litres of water was required for a single pair of jeans to complete its entire life cycle, with around 45% of the water being spent during the use phase.[23] The major hot spots in the life cycle of jeans were cotton production consuming 1704 litres followed by the use phase, which demanded 1575.2 litres. Other phases, namely, fabric production, clothing production, logistics and end of life, consumed 72.1, 110.8, 18.1 and 0.4 litres, respectively.[24]

There is another case study on calculating the WF of a pair of jeans, published in 2013. This study assessed the WF of five types of textiles used commonly for the production of jeans, including two different fibres, namely, cotton and lyocell and five textile production processes for spinning, dyeing and weaving. As expected reasonably, the results from this study have demonstrated that fibre production is the stage with the highest water consumption, in particular to cotton production. As per the results deduced from the study, total WF of cotton trousers on an average was 3233 (m^3/item) and, on the other hand, total WF of lyocell trousers on an average was 1454 (m^3/item). In terms of fibre production, for the cotton trousers the process of ginning contributed to only blue WFs and grey and green WFs were completely not there. In case of lyocell, it was the other way around, only grey WF was applicable and almost blue and green WFs were not there. For spinning and weaving, only blue WF was relevant for trousers made from both the fibres chosen for this study.[25]

There were many discussions and conclusions, and out of them, one of the major points to be noted is cellulose-based lyocell fibre has a notably lower water consumption than cotton fibres, on average, with 1384, 34.5 and 35.3 m^3/t as opposed to 263, 2767 and 203 m^3/t for green, blue and grey water, respectively.[25]

Another recent study conducted in 2017 measured the WF of viscose and polyester and compared it against that of cotton. According to the results from 15 production locations chosen for this study, WF of polyester staple fibres ranged from a minimum of 51,066 m^3/t to a maximum of 71,409 m^3/t and the WF of polyester filament yarns ranged from a minimum of 50,690 m^3/t to a maximum of 71,033 m^3/t. This is mainly from blue and grey WFs, as polyester does not have a green WF and the proportion of blue in the total WF is almost nil or very minimum compared with grey WF. For instance, in case of polyester staple fibres, 71,377 m^3/t was from grey and 32 m^3/t was from blue to make out to a maximum WF of 71,409 m^3/t.[26]

In case of viscose, viscose staple fibres, wood from plantations in Brazil was accounted for a WF of 678 m^3/ton, whereas wood from plantations in South Africa accounted for a WF of 996 m^3/t. Viscose filament yarn from continuous washing resulted in 30,596 m^3/t for Brazilian wood and 30,914 m^3/ton for South African wood. Viscose filament yarn from batch washing was found to have 3305 m^3/t for Brazilian wood and 3624 m^3/t for South African wood. For viscose, grey WF was predominantly higher than blue and green WFs. Again green WF was found to be very less than blue WF.[26]

Results from this study were compared to the results of other two studies on cotton. A detailed comparison was not made, as the data and scope limitations of this study under discussion and the ones chosen on cotton restricted the accuracy of the comparisons. Alongside the limitations noted already in view of comparisons, some important highlights between the WFs of three fibres considered under this study are (1) on an average, polyester has the highest WF (rarely, cotton surpassed this result from some conventional cotton farms in India, in which highly toxic pesticides are used); (2) viscose has the lowest WF (with some noted exceptions such as filament yarns produced with continuous washing, which has higher footprint for cotton from farms using REEL or organic farming practices, as well as cotton's global average WF); and—(3) the grey WF contributed a higher proportion of the total WF for viscose and polyester than it did for cotton (exceptions are farms that use highly toxic pesticides).[26]

4.5 Energy footprints: introduction and methods of assessment

After carbon and water, one of the most important indicators for measuring sustainability is energy. Energy is another resource becoming scarce globally, and the global energy crisis is an important topic deserving urgent action. There are many types of energy sources available today. Energy sources can be categorized into one of the three following types:

1. Fossil fuels such as coal, petroleum and natural gas.
2. Renewable energy sources such as solar, wind, hydroelectric, biomass and geothermal.

3. Nuclear energy sources such as nuclear fusion and nuclear fission.

The first category, fossil fuels, is the main source of global energy and poses the greatest challenge to our environment and society. More than 85% of the global energy supply is fulfilled by fossil fuels. Renewable and nuclear sources contribute only 7% and 6%, respectively.[27] According to the estimation of the US Department of Energy, the global energy demand is 400 quadrillion British thermal units (BTUs) per annum. Around 350 quadrillion BTUs is provided by fossil fuels, which is almost 90% (44% oil, 24% coal and 22% natural gas).[27]

Because of the over-consumption due to a spike in population, the world's energy demand is expected to hit 600 quadrillion BTUs in the next decade. There will be a quantum leap in the rate of depletion of fossil fuels to supply this energy demand if we rely only on fossil fuels. The US Department of Energy has predicted that the entire supply of fossil fuels will have been exhausted before the twenty-second century, if we rely solely on them.[27] This makes clear the need to develop alternative energy sources such as renewable energy.

Apart from its depletion effect, burning fossil fuels to generate electricity poses numerous other environmental considerations such as severe air pollution and increased global warming. Hence, there is an urgent need to measure energy consumption of different products, processes and businesses and seek possible ways to reduce it. The energy footprint is one way to measure that energy consumption.

Many definitions for energy footprints exist and some are listed in the following.

- As defined by the Global Footprint Network, energy footprint is the sum of all those areas required to provide feed and non-feed energy.[28]
- According to the WWF (World Wide Fund for Nature), energy footprint is defined as the area needed to sustain energy consumption and is measured as the area of forest that would be required to absorb the resulting CO_2 emissions, excluding the proportion absorbed by the oceans.[29]
- Another definition given by Palmer in 1998 is a measurement of the land required to absorb those CO_2 emissions originating from the usage of energy.[30] This definition focuses on the resultant effect of energy use, i.e. CO_2 emissions, in order to highlight the issue and promote ways of reducing the same.[31]
- Schindler provides a further new definition, which is that the energy footprint corresponds to the cumulative energy demand for non-renewable energy resources, i.e. fossil and nuclear energy.[32]

Energy footprint can be measured in terms of local or global hectares and again, in units of energy/functional unit.[28,33] Energy footprint also has many sub-footprints such as fossil or fossil energy footprint, solar energy footprint, wind energy footprint and renewable energy footprint.[33] However, as the focus of this chapter is energy use or energy consumption in the textile and clothing supply chain, the following discussion is confined to the amount of energy demanded by textile and clothing products.

As with water use, energy use or consumption of energy can be quantified for a particular process or different processes in a supply chain or for a total product by tracing the direct and indirect energy needs of a particular process (cumulative energy needs can also be assessed for multiple processes). This can be expressed as energy use in megajoules (MJ) or any other relevant unit for a particular unit of production.

The energy footprint or energy consumption of a product can also be calculated by the LCA method, tracing the amount of energy used in the entire product life cycle from cradle to grave stages. This method accounts for the amount of different forms of energy (including feedstock energy) required to produce a product from the raw material extraction stage to the disposal stage. This approach, similar to water use, includes the energy use in the manufacturing processes of the product and its associated components, background processes such as extraction of raw materials, and generation of electricity. It also takes into account the energy use in other life cycle phases.

4.6 Applications of energy footprint assessment to key stages in the textile supply chain

Apart from being lengthy, the textile and clothing supply chain is energy intensive. Every stage or process in the supply chain of textiles and every phase in the life cycle of textile products demands large amounts of energy. The textile supply chain and textile products require energy (in various forms such as electricity, hot oil and steam)

- to operate various machines involved in textile manufacturing;
- to transport between various processes;
- to produce raw materials, accessories, chemicals and other necessary elements;
- to raise the temperature of water baths employed in dyeing and other wet processing operations;
- to operate heating and cooling systems;
- to operate lighting, humidification plants and office equipment.

The contribution of the textile industry in the total energy consumption of different manufacturing industries in a country is decided by the structure of the manufacturing sector in that particular country.[34] About 4% of the total energy use in China comes from the textile sector,[35] whereas in the United States the textile industry accounts for less than 2% of the total energy use.[36]

The textile industry requires electricity and fuel for its supply chain. Electricity is primarily used in operating various machines and fuels are employed mainly in the wet processing of textiles. The manufacturing census data acquired from the United States in 2002 reveals that 61% of fuel energy and 39% of electricity were employed in the US textile industry. In the United States, the textile industry ranked as the fifth largest steam-consuming sector among 16 major industrial sectors.[37]

4.6.1 Energy consumption in fibre manufacture

There are numerous types of textile fibres used in producing clothing products. Every fibre has its own energy requirements from the extraction stage to manufacture. There is no one study that deals with the quantification of energy needs of different textile fibres. If such a study were available, then it would be possible to compare various fibres in terms of their energy needs. However, as such a study is not currently available, the energy needed to produce 1 kg of fibre is taken from various references and tabulated in Table 4.1. As each study is dependent on different factors such as its boundaries, assumptions and geographical location of the fibre manufacture the results from various studies will be different.

The results tabulated in Table 4.1 are for both natural and synthetic fibres. For natural fibres, it shows the amount of energy needed to produce a particular fibre in a mill (field to mill gate stages), and as far as synthetic fibres are concerned, it shows the energy required from raw material extraction to polymerization stage (that is, until the conversion to spinnable fibre stage).

4.6.2 Energy consumption in spinning, weaving and wet processes

There are many processes and machines involved in converting a fibre to yarn. Again, the process of spinning varies between natural and synthetic fibres as discussed in Chapter 1. There are many methods or mechanisms for spinning, such as ring spinning, rotor spinning, air-jet spinning, friction spinning and air vortex spinning. Each method has unique energy needs and its own dedicated process flow.

Ring and rotor (open-end) spinning techniques are quite popular for spinning cotton and cotton blends. The energy requirement per 1 kilogram of ring-spun yarn varies between 3.49 and 3.62 kW h/kg, whereas in the case of open-end spinning, energy requirements vary between 2.44 and 2.58 kW h/kg of yarn.[39] It should also be noted that energy consumption to produce a unit of yarn will vary significantly amongst various countries. For instance, ring- and rotor-spun yarns produced in China consume 3.49 and 2.58 kW h/kg, respectively, whereas in India and the United States the energy consumption is 3.57–2.5 kW h/kg and 3.56–2.44 kW h/kg, respectively.[39]

Table 4.1 Energy needs in fibre production

Fibre type	Energy consumption (MJ/kg of fibre)
Conventional cotton	60[21]
Organic cotton	54[21]
Nylon 66	138.65[22]
Nylon 6	120.47[22]
Viscose	100[38]
Polypropylene	115[38]
Polyester	125[38]
Acrylic	175[38]

Finer yarns require more energy than coarser ones and this is applicable to all types of fibre. There is also a distinction in the energy consumption between yarns used for weaving and knitting. For the same count, yarns for weaving consume more energy than those used for knitting. For example, 100 kg of 20 Tex yarn to be used in knitting consumes 306 kW h, whereas the same amount of yarn consumes 364 kW h if it is to be used for weaving.[40]

A spinning plant utilizes 78% of energy in its machines, 16% for humidification plant, 3% for lighting and 3% for compressors. A spinning plant consisting of both ring and rotor spinning machines uses 11% energy for blow room, 12% for carding, 5% for drawing, 1% for combing, 7% for roving, 37% for ring spinning machines, 20% for open-end machines and 7% for winding machines.[39]

The process of weaving comprises many sub-processes, namely, winding, warping, sizing, weaving preparation and weaving itself. From a study by Visvanathan and colleagues,[41] it was revealed that 1 kg of woven fabric consumes between 5.7 and 5.8 kW h of specific electrical energy. Energy consumption of weaving processes was measured by a study, which found that automatic winding needs 0.4 kW h/kg, classical winding and warping each need 0.1 kW h/kg and sizing consumes 0.03 kW h/kg of electrical energy and 5 MJ/kg of thermal energy.[42] In a weaving mill, 36.3% of energy is required for the machines, 29.4% for compressors, 27.1% for air conditioners, 3.4% for illumination and 3.8% for cleaning.[43]

Wet processing is the major component in the textile supply chain in terms of energy consumption because it demands huge amounts of thermal energy in steam and heat. The energy used in wet processing differs between different product forms, machine types, processes, process flows and state of the final product.[31] For instance, a desizing unit consumes 1.0−3.5 GJ/t of output, scouring by J-box consumes 6.5−10.0 GJ/t of output and bleaching by jig/winch requires 3.0−6.5 GJ/t of output. The amount of energy required for dyeing varies depending on which type of machine is employed. Dyeing by winch consumes 6.0−17.0 GJ/t of output and by jet dyeing machine requires 3.5−16.0 GJ/t of output. Printing by rotary screen consumes 2.5−8.5 GJ/t of output. Stenter needs 2.5−7.5 GJ/t of output for drying and 4.0−9.0 GJ/t of output for heat setting.[44]

In a composite textile plant, the ring spinning process consumes 41% of energy, weaving preparation and weaving consume 5% and 13%, respectively, humidification consumes 19%, wet processing consumes 10%, lighting demands 4% and other elements require 8%. This energy breakdown reveals that ring spinning is the major energy consumer in a composite textile plant. As far as thermal energy is concerned, bleaching and finishing require 35%, dyeing and printing consume 15%, humidification and sizing demand 15% and boiler plant and steam distribution losses account for 25% and 10%, respectively.[34]

Energy use in terms of the life cycle of a textile product deserves significant consideration and it varies widely for different textile products. According to the LCA (cradle to grave) study conducted by Levi Strauss and Company in 2007, around 58% (226.6 MJ) of the total energy consumed in the entire life cycle of a pair of jeans was utilized in the use phase. The fabric manufacturing phase consumed 21%, which works out at 84.9 MJ, followed by the cut/sew/finish phase, which consumed 10%

(40.8 MJ). The logistics/retail phase accounted for 7% (29.8 MJ) of total energy consumption and cotton production was responsible for 4% (17.7 MJ). The end-of-life phase consumed much less energy, only 0.4 MJ.[24]

The same study also studied the life cycle of a pair of khaki trousers. Again, the use phase was the major hot spot responsible for 63% of total energy consumption or 231.6 MJ. Fabric production came next for this product, too, and accounted for 26% (96.9 MJ), followed by logistics/retail, which accounted for 7% (24.6 MJ) of the total life cycle energy. Cut/sew/finish and cotton contributed 2% each (7.7 and 6.7 MJ) in total life cycle energy consumption, whilst the end-of-life phase consumed a very small amount of energy, 0.2 MJ.[24]

Another study, conducted in 2011, dealt with the life cycle of a polyester/cotton shirt. This study investigated 227 g of a 65%/35% polyester/cotton shirt, which was laundered 52 times and discarded in a landfill. According to this study, the product in question consumed 102 MJ over its entire life cycle. The use phase accounted for 64% (65 MJ) and the shirt manufacturing process accounted for 36% (37 MJ) of the total energy. The disposal phase consumed only 0.02% of the total life cycle energy.[45]

A study dealing with the present and future sustainability of textiles in the United Kingdom presented case studies of three products. This study demonstrates the variations in energy profiles of different textile products in their cradle to grave assessments. Published in 2006, it dealt with the energy profile of three products over their entire life cycle—a cotton T-shirt, a viscose blouse and a nylon carpet. A total of 109 MJ of energy was consumed in the entire life cycle of a single cotton T-shirt. The use phase, evaluated over 25 washes at 60 °C followed by drying and ironing, was found to be the major component requiring 65 MJ of energy. Production and material came next requiring 24 and 16 MJ, respectively, followed by transportation, which consumed 7 MJ. It was assumed that the T-shirt was incinerated and showed a negative energy value (−3 MJ) per piece.[46]

The viscose blouse had a different energy profile compared with that of the cotton T-shirt. Here the material phase was found to be the major hot spot, with 65% of the total energy consumption (33 MJ per piece) followed by production, use and transportation stages with 11, 7 and 3 MJ of energy consumed per piece, respectively. The use phase, not such an energy-intensive phase, was also evaluated over 25 washes but at 40 °C, requiring hang drying and no ironing. The disposal phase was the same as for the cotton T-shirt and consumed the same amount of energy (−3 MJ).[46]

As far as the carpet was concerned, it was made out of nylon fibres with polypropylene and latex-foam backing. Here the material production was found to be the major hot spot, consuming 275 MJ/m^2 of carpet (71% of the total energy). The use phase, assuming a 10-year life span and cleaned by vacuum cleaning, was found to be the second major hot spot, consuming 65 MJ of energy. Production and transportation consumed 40 and 10 MJ of energy, respectively. The disposal phase consumed no energy at all.[46]

This study showed that the energy consumed in material production varies depending on the type of material and is based on the product and its function. The energy profiles of production and use phases also vary. The method and temperatures of washing and drying based on the material type, life time and end use of the product affect the energy consumption in the use phase, and the energy consumption in the transportation phase varies according to the weight of product, distance to be transported and the means of transportation employed. Energy consumption in the disposal phase depends primarily upon the disposal option selected.

4.7 Summary: key challenges in calculating water and energy footprints

Following carbon, conservation of water and energy consumption for any product/process/business deserves significant attention. The measurement and conservation of water and energy consumption in any industrial supply chain needs to be addressed urgently. The textile and clothing supply chain consumes both water and energy in huge amounts so it is imperative to measure the energy footprint and WF of any textile product's entire life cycle stages in order to work on reducing them. As with carbon footprints, disclosure of energy footprint and WF is also becoming very important, although it has not yet become mandatory. Energy footprint and WF are also in the list of primary indicators to be displayed with the environmental impact information on products.

Similar difficulties to carbon footprint accounting of the textiles and clothing supply chain apply to the measurement of its WF and energy footprint. Some of the main difficulties in the modelling and simulation of WF and energy footprint across the supply chain of textiles and clothing are

- selecting a proper methodology for the study;
- defining and finalizing boundaries;
- goal and scope of the study;
- dealing with the choices of data and data quality;
- solving issues pertaining to data uncertainty;
- technical know-how difficulties in data collection and processing;
- filling data gaps with appropriate data;
- availability of suitable life cycle inventories to measure energy footprint and WF;
- selection and availability of appropriate secondary data to fill any data gaps;
- defining assumption of studies;
- tackling methodological complexities, such as allocation of products and co-products;
- solving the issues of cut-off rules of the study;
- identifying suitable products for the study from the long list available;
- availability of suitable databases for textiles and clothing products and their life cycle phases;
- verifying and validating results, if not done by an approved third party.

A typical issue in the textile sector is that the product profiles of calculated products keep changing resulting in the need to reassess the energy footprint and WF calculation. As said earlier, it is not possible to compare various textile product forms or complete products in terms of energy and water consumed in the entire life cycle stages because each study is unique in terms of methodology, geographical location in which the product was made, inventories chosen, assumptions and boundaries. In the same way, it is difficult to compare the energy footprint and WF of various textile products on the shelf.

4.8 Sources of further information and advice

- Aldaya MM, Martinez-Santos P, Llamas MR. Incorporating the water footprint and virtual water into policy: reflections from the Mancha Occidental Region, Spain. *Water Resources Management.* 2010;24(5):941–958.
- Allan JA. Water security policies and global systems for water scarce regions. In: *Sustainability of Irrigated Agriculture—Transactions, Vol. 1E, Special Session: The Future of Irrigation under Increased Demand From Competitive Uses of Water and Greater Needs for Food Supply–R. 7 in the Symposium on Management Information Systems in Irrigation and Drainage, 16th Congress on Irrigation and Drainage, Cairo.* New Delhi: International Commission on Irrigation and Drainage; 1996.
- Allan JA. Virtual water–the water, food, and trade nexus: useful concept or misleading metaphor? *Water International.* 2003;28(1):106–113.
- Bos E, Bergkamp G. Water and the environment. In: Meinzen-Dick RS, Rosegrant MW, editors. *Overcoming Water Scarcity and Quality Constraints, 2020 Focus 9.* Washington, D.C.: International Food Policy Research Institute; 2001.
- Chapagain AK, Hoekstra AY. The global component of freshwater demand and supply: an assessment of virtual water flows between nations as a result of trade in agricultural and industrial products. *Water International.* 2008;33(1):19–32.
- Chapagain AK, Hoekstra AY. *The Green, Blue and Grey Water Footprint of Rice from Both a Production And Consumption Perspective.* Value of Water Research Report Series No. 40, Delft, Netherlands: UNESCO-IHE; 2010.
- Dyson M, Bergkamp G, Scanlon J, editors. *Flow: The Essentials of Environmental Flows.* Gland, Switzerland: IUCN; 2003.
- Falkenmark M, Rockström J. *Balancing Water for Humans and Nature: The New Approach in Ecohydrology.* London: Earthscan; 2004.
- Hoekstra AY. *The Relation Between International Trade and Freshwater Scarcity.* Staff Working Paper ERSD-2010-05. World Trade Organization, Economic Research and Statistics Division; January 2010.
- Mila i Canals L, Chenoweth J, Chapagain A, Orr S, Anton A, Clift R. Assessing freshwater use in LCA: Part I—inventory modelling and characterisation factors for the main impact pathways. *International Journal of Life Cycle Assessment.* 2008;14:28–42.
- Seckler D, Amarasinghe U, David M, de Silva R, Barker R. *World Water Demand and Supply, 1990 to 2025: Scenarios and Issues.* Research Report 19. Colombo, Sri Lanka: International Water Management Institute; 1998.
- Shiklomanov IA. Appraisal and assessment of world water resources. *Water International.* 2000;25(1):11–32.
- Smakhtin V, Revenga C, Döll P. A pilot global assessment of environmental water requirements and scarcity. *Water International.* 2004;29:307–317.
- United Nations World Water Assessment Programme. The United Nations World Water Development Report 3—Water in a Changing World; UNESCO: Paris, France; 2009.

4.9 References

1. Casella C. *Nearly 25% of the World's Population Faces a Water Crisis, and We Can't Ignore It.* Available at: https://www.sciencealert.com/17-countries-are-facing-extreme-water-stress-and-they-hold-a-quarter-of-the-world-s-population [accessed on 1st April 2019].
2. *Projected Water Scarcity in 2025.* Available at: http://peakwater.org/2010/02/projected-water-scarcity-in-2025/ [accessed on 10 April 2013].
3. Rosegrant MW, Cai X, Cline SA. *Global Water Outlook to 2025.* Sri Lanka: International Food Policy Research Institute, USA and International Water Management Institute; September 2002.
4. Chapagain AK, Hoekstra AY, Savenije HHG, Gautam R. The water footprint of cotton consumption: an assessment of the impact of worldwide consumption of cotton products on the water resources in the cotton producing countries. *Ecological Economics.* 2006;60(1): 186–203.
5. Chapagain AK, Hoekstra AY. Virtual water flows between nations in relation to trade in livestock and livestock products. *Value of Water Research Report Series.* 2003;13. Delft, the Netherlands: UNESCO-IHE.
6. Hoekstra AY, Chapagain AK. Water footprints of nations: water use by people as a function of their consumption pattern. *Water Resources Management.* 2007;21(1):35–48.
7. Hoekstra AY, Chapagain AK, Aldaya MM, Mekonnen MM. *Water Footprint Manual: State of the Art 2009.* Enschede, the Netherlands: Water Footprint Network; 2009.
8. Galli A, Wiedmann T, Ercin E, Knoblauch D, Ewing B, Giljum S. *Integrating Ecological, Carbon and Water Footprint: Defining the – Footprint Family and its Application in Tracking Human Pressure on the Planet.* Technical Document. Open EU; January 28, 2011. Available at: http://www.oneplaneteconomynetwork.org/resources/programme-documents/WP8_Integrating_Ecological_Carbon_Water_Footprint.pdf.
9. *Glossary of Water Footprint Terms.* Available at: http://www.waterfootprint.org/?page=files/Glossary.
10. *The Concepts of Water Footprint and Virtual Water.* Available at: http://www.gdrc.org/uem/footprints/water-footprint.html.
11. Allan JA. Virtual water: a strategic resource, global solutions to regional deficits. *Ground Water.* 1998;36(4):545–546.
12. *Water Footprint – Concepts and Definitions.* Available at: http://www.gracelinks.org/1336/water-footprint-concepts-and-definitions.
13. Hoekstra AY. Human appropriation of natural capital: a comparison of ecological footprint and water footprint analysis. *Ecological Economics.* 2009;68:1963–1974.
14. Hoekstra AY, Chapagain AK, Aldaya MM, Mekonnen MM. *The Water Footprint Assessment Manual: Setting the Global Standard.* Earthscan; 2011. ISBN 9781849712798.
15. Berger M, Finkbeiner M. Water footprinting: how to address water use in life cycle assessment? *Sustainability.* 2010;2:919–944.
16. *Water Footprint of Different Products from Product Gallery.* Available at: http://www.waterfootprint.org/?page=files/productgallery&product=cotton.
17. Mekonnen MM, Hoekstra AY. The green, blue and grey water footprint of crops and derived crop products. In: *Value of Water Research Report Series No. 47.* UNESCO-IHE; 2010.
18. Mekonnen MM, Hoekstra AY. The green, blue and grey water footprint of crops and derived crop products. *Hydrology and Earth System Sciences.* 2011;15(5):1577–1600.

19. Janet Bealer Rodie. *Fiber First, Textile World*; 2011. September/October Issue. Available at: http://www.textileworld.com/Articles/2011/September/Sept_Oct_issue/Sustainability_Fiber_First.html.
20. Cherrett N, Barrett J, Clemett A, Chadwick M, Chadwick MJ. *Ecological Footprint and Water Analysis of Cotton Hemp and Polyester*. Stockholm, Sweden: Stockholm Environment Institute; 2005.
21. Kaillala ME, Nousiainen P. Environmental profile of cotton and polyester cotton fabrics. *Autex Research Journal*. 1999;1(1):8–20.
22. Boustead I. *Eco-profiles of the European Plastics Industry*. Brussels: Plastics Europe; 2005.
23. Water. Levi Strauss Website. Available at: http://www.levistrauss.com/sustainability/planet/water.
24. *A Product Life Cycle Approach to Sustainability*. Available at: http://www.levistrauss.com/sites/default/files/librarydocument/2010/4/Product_Lifecyle_Assessment.pdf.
25. Daniel CZ, Aldaya MM, Alberto GC. A water footprint assessment of a pair of jeans: the influence of agricultural policies on the sustainability of consumer products. *Journal of Cleaner Production*. 2013;57:238–248.
26. Freitas A, Zhang G, Mathews R. *Water Footprint Assessment of Polyester and Viscose and Comparison to Cotton*. Water Footprint Network; March 2017.
27. *Renewable Energy Vs Fossil Fuel*. Available at: http://www.renewablepowernews.com/archives/1413.
28. *Global Footprint Network, 2009, Glossary*. Available at: www.footprintnetwork.org.
29. WWF (World Wide Fund for Nature). *Living Planet Report*; 2002. Available at: www.wwf.de.
30. Palmer ARP. Evaluating ecological footprints. *Electronic Green Journal*. 1998;1(9). Available at: escholarship.org/uc/item/05k183c9.
31. *What is Energy Footprint. Eco-Footprints Page*. Available at: http://www.gdrc.org/uem/footprints/energy-footprint.html.
32. *Schindler. Energy & GHG Footprint*. Available at: http://www.schindler.com/com/internet/en/about-schindler/corporate-citizenship/site-ecology/energy-and-ghg-footprint.html.
33. Cucek L, Klemes JJ, Kravanja Z. A review of footprint analysis tools for monitoring impacts on sustainability. *Journal of Cleaner Production*. 2012;34:9–20.
34. Hasanbeigi A, Price L. A review of energy use and energy efficiency technologies for the textile industry. *Renewable and Sustainable Energy Reviews*. 2012;16(6):3648–3665.
35. Jones T. Steam partnership: improving team efficiency through market place partnerships. In: *Proc. 1997 ACEEE Summer Study on Energy Efficiency in Industry*. Washington DC, USA: ACEEE; 1997.
36. United States Department of Energy (US DOE). *Industrial Assessment Center (IAC), Industrial Assessment Center (IAC) Database*. Washington, DC, USA: Department of Energy; 2006. Available at: http://iac.rutgers.edu/database/index.php.
37. United States Department of Energy (US DOE). *Improving Ventilation System Energy Efficiency in a Textile Plant*; 2005. Available at: http://www1.eere.energy.gov/industry/bestpractices/case_ study_ ventilation_textile.html.
38. Barber A, Pellow G. *Life Cycle Assessment New Zealand Merino Industry, Merino Wool, Total Energy Use and Carbon Dioxide Emissions*. Pukekohe, Auckland: The Agri Business Group; 2006.
39. Koc E, Kaplan E. An investigation on energy consumption in yarn production with special reference to ring spinning. *Fibres and Textiles in Eastern Europe*. 2007;15(63):4.
40. Lanham G. 4 factors to lower energy costs. *Metal Producing and Processing*. 2007;45(5):8.

41. Visvanathan C, Kumar S, Han S. *'Cleaner Production in Textile Sector: Asian Scenario', National Workshop on 'Sustainable Industrial Development through Cleaner Production', Sri Lanka*; 2000. Available at: http://www.serd.ait.ac.th/smi2/SMI/roadmap/Presentations/CP_Textile_Asia.pdf.
42. Senol F. *Energy Consumption and Conservation in Weaving Preparation Mill, Seminar Notes about Energy Consumption and Conservation in Textile Mills*. Bursa: Sümerbank Textile Education and Research Institute; 1985. Number 77 (in Turkish).
43. Koc E, Cincik E. Analysis of energy consumption in woven fabric production. *Fibres and Textiles in Eastern Europe*. 2010;18(2):14–20.
44. *Carbon Trust Low Cost Heat Recovery at W & J Knox Ltd, Ayrshire. Good Practice*; 2005. Available at: www.carbontrust.co.uk.
45. Cartwright J, Cheng J, Hagan J, Murphy C, Stern N, Williams J. *Assessing the Environmental Impacts of Industrial Laundering: Life Cycle Assessment of Polyester/cotton Shirts*. Donald Bren School of Environmental Science & Management, University of California; April 2011. Available at: http://www.bren.ucsb.edu/research/documents/missionlinen_brief.pdf.
46. Allwood JM, Bocken N, Laursen SE, Malvido de Rodriguez C. *Well Dressed? The Present & Future Sustainability of Clothing & Textiles in the UK*. Cambridge: University of Cambridge, Sustainable Manufacturing Group, Institute for Manufacturing; 2006.
47. Hoekstra AY, Hung PQ. Virtual water trade: a quantification of virtual water flows between nations in relation to international crop trade. *Value of Water Research Report Series*. 2002; 11. Delft, the Netherlands: UNESCO-IHE.
48. USEPA. *Best Management Practices for Pollution Prevention in the Textile Industry*; 1996. www.e-textile.org.

Textile processing and resource depletion: calculating the ecological footprint of textile products

5

5.1 Introduction

A basic issue for environmental resources is the regenerative capacity of natural resources. Consumption of natural resources should not exceed available reserves or their capacity to renew themselves, e.g. for rivers to replenish water extracted from them. However, economic growth and over-consumption means that natural resources are being used and depleted in an unsustainable way. It is critical to have some measure of sustainability. The concept of 'ecological footprint' (EF) has been developed to quantify how sustainable a product or process is.

The EF is a sustainability indicator along with the carbon, water and energy footprints and it defines the biologically productive land area required to support the current consumption pattern of the global population. Measurement of the EF of textiles and the clothing supply chain defines the degree of sustainability for these activities from fibre production (including cultivation and extraction) through to garment manufacturing, use and disposal. This chapter concerns the concept of the EF, its principles and measurement, with specific reference to textile processes and products.

5.2 Main principles and methods

The increasing global population is placing unsustainable demands on the environment, as consumption exceeds the capacity of eco-systems to regenerate and renew the resources required for survival, and waste is generated at a rate that cannot be easily accommodated.[1] This obviates the need to account for the resources depleted by a product or process, a person or a country as a whole and is achieved through a methodological principle called the 'ecological footprint'. This attempts to quantify the conditions required to achieve sustainability through assessment of the biological area required to sustain current resource-depleting activities.[2] We can thus express the sustainability of human activity through an area measurement that is easily understood and communicated.[3]

The concept of EF was first introduced by Mathis Wackernagel and William Rees of the University of British Columbia,[4] who developed a resource-accounting tool that computes how much biologically productive land and water area an individual, a city, a country, a region or the entire global population requires in order to maintain its

Assessing the Environmental Impact of Textiles and the Clothing Supply Chain
https://doi.org/10.1016/B978-0-12-819783-7.00005-3

current consumption, and for the environment to absorb its generated waste. The result is commonly expressed in terms of global hectares.

As defined in the 2008 edition of the *National Footprint Account*, EF takes into account the following six categories:

1. cropland,
2. forest land (relating to both timber and carbon sequestration purposes),
3. grazing land,
4. fishing ground,
5. energy footprint,
6. built-up land.

The footprint for any of these six categories represents the area required to produce the products or goods representative of that particular area. The sum of these values represents the value of the global footprint.[5]

The EF of each major land-use type, except built-up land and forests for carbon sequestration, is calculated by summing the contributions of a specific product or a variety of products. Built-up land is an indicator of bio-productivity, which is compromised by roads and urban areas. Forests needed to sequester anthropomorphic carbon dioxide represent the required capacity to absorb this particular type of waste. The calculation of EF spins around two major issues covered in the six land types indicated earlier, which are demand for resource production and waste assimilation. These two major issues are translated into global hectares by the values of the following parameters:

- total amount of a resource consumed/yield per hectare,
- waste emitted/absorptive capacity per hectare.

The EF of any consumption activity is calculated according to

$$\text{Ecological footprint} = (\text{Annual demand in tonnes / National yield in annual tonnes per hectare}) \times \text{Yield factor} \times \text{Equivalence factor} \quad [5.1]$$

The yield factor forms a useful comparison of the national average yield per hectare to the world average yield in the same land category. The equivalence factor defines the relative productivity among the land and sea area types.

5.2.1 Ecological footprint accounting of products and processes

By means of the depletion or use of resources, generation and assimilation and the utilization of services from nature, every individual, process, product, action and region impacts our living planet. These impacts must be accounted for and should be converted to biologically productive areas. The EF accounts for the human appropriation of ecological production in area units and expresses the exploitation of natural resources by human activities in a single dimension of land and water demanded to support them.[6]

The EF can be calculated for an individual, a village, a locality, a nation or the whole world itself, along with organizations, services, products and processes.[7] Similar to other indicators discussed in previous chapters, EF can be calculated for different products by following a life cycle approach and can also be calculated for various processes to produce a particular product.

The EF of a product can be measured for cradle to gate or cradle to grave stages or at any stage associated with the life cycle assessment (LCA) of that product. By mapping the entire life cycle phases associated with the chosen product, one can quantify its EF through consideration of three parameters:

1. The amount of area required for the production of crops, forest and animal products relating to that product.
2. The area needed to sequester atmospheric carbon dioxide emissions, principally related to the combustion processes of fossil fuels used in the generation of the product.
3. The area required for producing the nuclear energy for the generation of the product.[8–10]

From an LCA perspective, the EF of a product can be defined as the sum of the time-integrated direct and indirect land occupation, where the latter is related to nuclear energy use and to CO_2 emissions from fossil fuel use. It is calculated as[10,11]

$$EF = EF\ (\text{direct}) + EF\ (CO_2) + EF\ (\text{nuclear}) \qquad [5.2]$$

The EF of a finished product can be calculated via one of the two approaches, namely, process-based LCA and environmentally extended input–output LCA. With the former method, however, complete upstream coverage of the production chain is not possible.[12]

The EF can be calculated for any products or processes in accordance with ISO 14040 and 14044 standards by following the four phases of LCA defined by these standards. Having clearly defined the goal and scope of the study with earmarked boundaries, a functional unit of the assessment has to be defined. Then the usual steps of LCA will follow, namely, collection of life cycle inventory, EF assessment (life cycle impact assessment) and interpretation of results.

5.3 Application to key stages in the supply chain

As discussed earlier, the EF is one of the important indicators of ecological sustainability used in sustainability assessment and LCA. Applicable to any industry, it portrays the amount of area on Earth required to maintain our consumptive characteristics and to digest the waste generated. The textiles and clothing supply chain demand a lot of resources in the form of energy, land and water, in addition to various material and chemical requirements. Each stage of the life cycle chain for textile products contributes to the EF, from fibre production to waste disposal (i.e. cradle to grave). Fibre production is in fact a major contributor to the EF for the clothing supply chain. As far as cotton is concerned, water needs and chemicals usage in the growth stage of fibre (especially for conventional cotton) increase the EF of cotton and also the final product made out of it. However, a study conducted in 2012 revealed that African cotton produced under an initiative known as 'Cotton made in Africa' (CmiA) and founded by the Aid by Trade Foundation (AbTF) in 2005 resulted in a lower EF than conventional cotton. Although this study did not directly quantify the EF, it did quantify the carbon and water footprints of African cotton and compared them with those of its rivals, which essentially form a basis to understand the EF.[13]

The initiative's sustainable cultivation methods reduce greenhouse gas emissions by over 70% and save around 18,000 litres of water per kilogram of cotton lint compared with Pakistani cotton. Rainfed cultivation methods are utilized for the growth of CmiA. Best ecological gains are achieved in terms of CmiA cotton with zero m^3 of fresh water per kilogram of cotton lint vis-à-vis the average conventional cotton with an index of 5 m^3 per kilogram of cotton.[13,14]

According to this study, CmiA cotton has a carbon footprint of 1.92 kg CO_2-eq compared with conventional cotton that has a carbon footprint of 4.64 kg CO_2-eq. Major contributors to the carbon footprint of CmiA-produced cotton were mineral fertilizer production at 52%, N_2O fertilizer soil emissions at 17% and livestock emission at 12%. With conventional cotton, the similar breakdown was mechanical energy at 34%, fertilizer production at 33% and fertilizer soil emissions at 10%. The CmiA cotton is free from the emissions associated with mechanical energy, and the further education of farmers on sustainable measures of agriculture and efficient measures and practices would lower the footprint of CmiA cotton further. The water footprint of CmiA cotton (14.2 m^3, 99% green water) is slightly higher than that for conventional cotton (13.1 m^3, 40% green water). Owing to the absence of irrigation, CmiA cotton does not utilize blue water.[13]

A study conducted in 2005 by the Stockholm Environment Institute analysed the energy requirements, carbon dioxide emissions, EFs and water requirements of five major textile fibres, namely, conventional and organic cottons, conventional and organic hemps and polyester produced in different countries.[15] The EFs were determined in global hectares for the production of 1 tonne of spun fibre. From the results of this study, it was noted that polyester from Europe and the United States, with an EF of 1.67 and 2.21 gha, respectively, occupies a middle position between the two extremes of cotton and hemp. Hemp was found to have a lower EF between 1.46 and 2.01 gha when compared with cotton and polyester.[15]

Cotton and hemp require land for their fibre production, whereas polyester does not; however, the latter requires a huge amount of energy for its fibre production process. Crop cultivation for cotton and hemp was found to be the major contributor to the EFs in these cases. Organic cotton in the United States had an EF of 2.17 gha and Indian conventional cotton (grown in Punjab) had an EF of 3.57 gha. Although the energy requirements are lower for the Indian cotton system grown in Punjab, lower yields compared with those of the United States results in a higher overall EF.[15]

Another relevant study was conducted in Spain in 2008.[16] This study analysed the EF of a garment manufacturing plant. The factory evaluated manufactured jackets from cotton fabric, and the entire manufacturing process of the plant was included in the study. Fabrics arriving at this factory were cut and sewn according to the desired pattern, as were various accessories such as buttons and zips. The jacket was then ironed, labelled, packed and stored in a warehouse for later distribution. Energy, resources and waste were the three categories concerned, representing the consumption and generation patterns involved in the EF concept.

A detailed process inventory was calculated comprising all major inputs such as raw materials, energy requirements (electricity, wind power, propane, gas, oil), water requirements, production outputs, air emissions and different kinds of waste for the years 2002–05. The EF was calculated according to the approach by Herva and co-workers.[16] The results indicated that the EF increased continuously year-on-year and that resource consumption played a major role, followed by the energy and waste generation categories. In the year 2005, resources, energy consumed and waste generated contributed to around 91.3%, 5.3% and 3.4%, respectively, in the case of synthetic stitch and cotton jackets with no recycling category. Clearly, the notable contribution arises from the cotton textile and results from the land area required to produce it.

In the current study, synthetic stitch was replaced by wool and for the same year, i.e. 2005, and similar conditions the EF has a 93.4% contribution from resources and around 4% and 2.6% contributions from energy and waste generated, respectively. Notable variation in the results arises from the replacement of wool for the synthetic stitch. In this case, wool is responsible for a 25% contribution to the total EF, bringing down the contribution of cotton from 77.4% to 58.8%. This is mainly due to the requirement for pastureland in wool production and for arable land in cotton production.

According to Herva and colleagues,[16] the EF is greatly influenced by the materials employed in manufacturing a product, the energy sources deployed in the manufacturing process and the method of handling waste. A study conducted by the same authors in 2012 dealt with the combined energy and material flows for the same factory.[17] The authors found an increment in the EF results in this revised work because of the inclusion of other emissions in addition to CO_2, stressing the importance of including all emissions associated with the production processes.

Other recent studies have calculated the EF of the most commonly used materials in the United Kingdom, United States, Canada, Israel and Australia. Results indicate that paper, plastics, diapers, glass, metals and textiles were the six most commonly used material types from which 12 materials were selected. Of these, polyethylene terephthalate (PET), high-density polyethylene (HDPE), diapers and cotton textiles are discussed here.[18]

The results showed an EF, on average, of 0.20 gha per tonne of the material for HDPE, 0.48 gha for PET, 1.14 gha for diapers and 10.2 gha for cotton fabric. Clearly, cotton fabric was found to have the highest EF per unit of material mainly due to the demand for agricultural land and energy and other inputs such as water, fertilizers and other resources.[18]

5.4 Summary: key challenges in calculating ecological footprints in textiles

Sustainable development requires development and measurement of appropriate indicators. The EF for humanity is in need of such indicators, which can portray a clear and unambiguous message, with the aid of which humanity can proceed further in achieving the goal of attaining sustainable development. EF is one such indicator that can be applied to assess and compare the environmental sustainability of a wide array of goods and services. Results indicate that it will take 1.5 years for the Earth to regenerate the resources we consume, and assimilate the waste we generate, on an annual basis.[19] Furthermore, EF allows a link to be established between local and global consumption.[20] When it comes to textiles, as mentioned earlier, it is certainly one of the accurate and essential indicators to express the sustainability of textile products. The entire supply chain of textiles and clothing is well related to this indicator, although certain areas directly affect the EF. It is worth studying this indicator for various textile products and also for various specific elements in the supply chain individually. Although it is such an important indicator pertaining to textile products, the available studies measuring this indicator for textiles are scanty. As noticed in this chapter, only a few studies are available on measuring the EF of textile products. All the key challenges discussed in previous chapters, such as availability of databases and modelling difficulties, will also be applicable to the EF study and this could be the main reason for the scarcity of studies on the specific EF of textile products.

Although this is a quite variable and subjective indicator as other indicators dealt with in previous chapters, it is pivotal to have the information on the EF of various textile products (finished and even intermediate products) produced from different countries. This would be a great asset to the textiles and clothing sector but, if further research progresses, more information is needed on, but not limited to, the following aspects:

- the EF of various textile fibres grown in different parts of the world;
- the EF of different finished products made in various countries encompassing intermediate stages;
- the EF of various textile processes such as spinning, fabric formation, chemical processing and garmenting in different countries and for different fibres;
- the EF of similar textile products as used and disposed of in different countries;
- high contributing factors to the EF in the life cycle of various textiles and clothing products;
- the influence of consumption behaviour and functionality of textile products.

Such information would allow for planning on efficient means of reducing the EF of textile products.

5.5 Sources of further information and advice

- Best A, Giljum S, Simmons C, Blobel D, et al. *Potential of the Ecological Footprint for Monitoring Environmental Impacts From Natural Resource Use*. Report to the European Commission, DG Environment; 2008.
- Blasi E, Pasveri N, Franco S, Pancino B, Cicatiello C. Ecological footprint definition. In: *Academic Conference: The State of the Art in Ecological Footprint Theory and Applications*. Colle Val d'Elsa: Global Footprint Forum; 2010. pp. 17–18.

- Chambers N, Lewis K. *Ecological Footprint Analysis: Towards a Sustainability Indicator for Business, ACCA (Association of Chartered Certified Accountants).* Research Report No. 65. London, United Kingdom: Certified Accountants Educational Trust; 2001.
- Galli A, Kitzes J, Wermer, P, Wackernagel M, Niccolucci V, Tiezzi E. An exploration of the mathematics behind the ecological footprint. In: Brebbia C, editor. *Ecodynamics: The Prigogine Legacy.* Billerica, MA: Wit Press; 2007. pp. 249−256.
- Herva M, García-Diéguez C, Franco-Uría A, Roca E. New insights on ecological footprinting as environmental indicator for production processes. *Ecological Indicators.* 2012;16: 84−90.
- Holmberg J, Lundqvist U, Robert KH, Wackernagel M. The ecological footprint from a systems perspective of sustainability. *International Journal of Sustainable Development and World Ecology.* 1999;6(1):17−33.
- Kitzes J, Peller A, Goldfinger S, Wackernagel M. Current methods for calculating national ecological footprint accounts. *Science for Environment and Sustainable Society.* 2007;4: 1−9.
- Monfreda C, Wackernagel M, Deumling D. Establishing national natural capital accounts based on detailed ecological footprint and biological capacity assessments. *Land Use Policy* 2004;21(3):231−246.
- Nijkamp P, Rossi E, Vindigni G. Ecological footprints in plural: a meta-analytic comparison of empirical results. *Regional Studies.* 2004;38(7):747−765.
- Rees WE. Ecological footprints and appropriated carrying capacity: what urban economics leaves out. *Environment and Urbanization.* 1992;4(2):121−130.
- Senbel M, McDaniels T, Dowlatabadi H. The ecological footprint: a non-monetarymetric of human consumption applied to North America. *Global Environmental Change.* 2003;13: 83−100.
- Simmons C, Lewis K, Barrett J. Two feet − two approaches: a component-based model of ecological footprinting. *Ecological Economics.* 2000;32:375−380.
- Simmons C. *Further Development of the Ecological Footprint as an Indicator of Regional Sustainability;* 2002. European Commission Service Contract: B4-3-5-/2002/336545/MAR/B3.
- Wackernagel M, Yount JD. Footprints for sustainability: the next steps. *Environment, Development and Sustainability.* 2000;2:21−42.
- Wackernagel M, Schultz NB, Deumling D, Callejas Linares A, Jenkins M, et al. Humanity's resource demand exceeds the Earth's capacity. *Proceedings of the National Academy of Sciences.* 2002;99(14):9266−9271.
- Wackernagel, M., Monfreda, C., Moran, D., Wermer, P., Goldfinger, S., Deumling, D., and Murray, M. *National Footprint and Biocapacity Accounts 2005: The Underlying Calculation Method.* Oakland: Global Footprint Network; 2005
- Wiedmann T, Lenzen M. On the conversion between local and global hectares in ecological footprint analysis. *Ecological Economics.* 2007;60:673−677.
- Wiedmann T, Minx J, Barrett J, Wackernagel M. Allocating ecological footprints to final consumption categories with input-output analysis. *Ecological Economics.* 2006;56: 428−448.
- Wilson J, Tyedmers P, Pelot R. Contrasting and comparing sustainable development indicator metrics. *Ecological Indicators.* 2007;7:299−314.

5.6 References

1. Ewing B, Goldfinger S, Wackernagel M, Stechbart M, Rizk SM, et al. *The Ecological Footprint Atlas 2008*. Oakland: Global Footprint Network; 2008.
2. Holmberg J, Lundqvist U, Robèrt K-H, Wackernagel M. The ecological footprint from a systems perspective of sustainability. *The International Journal of Sustainable Development and World Ecology*. 1999;6:17–33.
3. Ruževičius J. Ecological footprint: evaluation methodology and international benchmarking. *Current Issues of Business and Law*. 2011;6(1):11–30.
4. Wackernagel M, Rees WE. *Our Ecological Footprint: Reducing Human Impact on the Earth*. Gabriola Island: New Society; 1996.
5. Jørgensen AE, Vigsøe D, Kristoffersen A, Rubin O. *Assessing the Ecological Footprint*. Environmental Assessment Institute; August 2002.
6. van den Bergh JCJM, Verbruggen H. Spatial sustainability, trade and indicators: an evaluation of the 'ecological footprint'. *Ecological Economics*. 1999;29:61–72.
7. Hoekstra AY. Human appropriation of natural capital: a comparison of ecological footprint and water footprint analysis. *Ecological Economics*. 2009;68:1963–1974.
8. Wackernagel M, Schulz NB, Deumling D, Linares AC, Jenkins M, et al. Tracking the ecological overshoot of the human economy. *Proceedings of the National Academy of Sciences of the United States of America*. 2002;99:9266–9271.
9. Monfreda C, Wackernagel M, Deumling D. Establishing national natural capital accounts based on detailed ecological footprint and biological capacity assessments. *Land Use Policy*. 2004;21(3):231–246.
10. Huijbregts MAJ, Hellweg S, Frischknechtc R, Hungerbühlerd K, Hendriks AJ. Ecological footprint accounting in the life cycle assessment of products. *Ecological Economics*. 2007; 64(4):798–807.
11. Frischknecht R, Jungbluth N, Althaus HJ, Doka G, Dones R, et al. *Implementation of Life Cycle Impact Assessment Methods: Data v2.0*. Ecoinvent Report No. 3. Dübendorf, Switzerland: Swiss Centre for Life Cycle Inventories; 2007.
12. Global Footprint Network. *Ecological Footprint Standards 2009*. Oakland: Global Footprint Network; 2009. Available at: www.footprintstandards.org.
13. Nill M, Wick K. *'The Carbon and Water Footprint of Cotton Made in Africa', Study Conducted for Aid by Trade Foundation*. Systain Consulting; October 2012.
14. *African Cotton Has a Small Ecological Footprint, Make Cotton Sustainable*, Sustainable Cottons.com. Available at: http://www.sustainablecottons.com/african-cotton-small-ecological-footprint.
15. Cherrett N, Barrett J, Clemett A, Chadwick M, Chadwick MJ. *Ecological Footprint and Water Analysis of Cotton Hemp and Polyester*. Sweden: Stockholm Environment Institute Stockholm; 2005.
16. Herva M, Franco A, Ferreiro S, Alvarez A, Rocaa E. An approach for the application of the Ecological Footprint as environmental indicator in the textile sector. *Journal of Hazardous Materials*. 2008;156:478–487.
17. Herva M, Alvarez A, Rocaa E. Combined application of energy and material flow analysis and ecological footprint for the environmental evaluation of a tailoring factory. *Journal of Hazardous Materials*. 2012;237–238:231–239.
18. Kissinger M, Sussman C, Moore J, Rees WE. Accounting for the ecological footprint of materials in consumer goods at the urban scale. *Sustainability*. 2013;5:1960–1973. https://doi.org/10.3390/su5051960.

19. Global Footprint Network. *Footprint Basics Overview*; 2012. Available at: http://www.footprintnetwork.org/en/index.php/GFN/page/footprint_basics_overview/.
20. Chambers, N. and Lewis, K. (2011). reportEcological Footprint Analysis: Towards a Sustainability Indicator for Business. ACCA Research Report No. 65, Certified Accountants Educational Trust, London.

Estimating the overall environmental impact of textile processing: life cycle assessment of textile products

6

6.1 Introduction

The assessment of the range of environmental impacts of different products on the planet is of growing importance and there are many techniques available to assess particular aspects of the environmental impact of a product. However, tracing and evaluating the environmental impact of a product during its entire life cycle from cradle to grave is of pivotal importance. Life cycle assessment (LCA) is the only method that evaluates the environmental impact of a product over its entire life cycle. It is a holistic approach that takes into account all the activities involved in the creation of a product, such as raw material extraction, manufacturing, transportation and distribution, use and disposal. It is therefore able to evaluate all the different potential types of environmental impacts associated with a product.

There are ISO standards relating to the LCA of any product, process or service. ISO standards 14040 and 14044 require inclusion of four stages, namely,

1. goal and scope definition
2. inventory analysis
3. impact assessment
4. interpretation.

Using LCA, one can understand the different environmental impacts associated with a product across its different life cycle phases. Most importantly, with the aid of LCA, one can figure out the hot spots in the entire life cycle of a product that are responsible for creating the most significant environmental impacts. This makes it possible to reduce those impacts by concentrating on problematic areas. Outcomes of LCA are used in diversified applications such as decision making, product development and engineering, or marketing.

The previous chapters have discussed different environmental impacts and their measurement. All techniques discussed so far have their roots in LCA, which is the key to measure any one environmental impact. This chapter discusses the history and basic principles of LCA, measurement methods and models for LCA and the standards governing an LCA study. It also discusses the key challenges of conducting an LCA study.

Assessing the Environmental Impact of Textiles and the Clothing Supply Chain
https://doi.org/10.1016/B978-0-12-819783-7.00006-5

6.2 History of life cycle assessment

LCA had its origins in the United States during the 1960s. Initially, focus was on energy efficiency, consumption of raw materials and, to a lesser extent, waste disposal.[1,2] Initial work was reported by Harold Smith at the World Energy Conference in 1963, which dealt with the calculation of energy needs for chemical products and their intermediates. Following this, in 1969, The Coca-Cola Company conducted an internal study to analyse different beverage containers in the context of resource consumption and other environmental impacts. This study focused on quantification of the raw materials employed, amount of fuel used and the environmental loadings due to manufacturing in order to identify the container that had the lowest environmental impact. Around the same timescale, other companies in Europe and the United States started performing similar life cycle inventory (LCI) analyses.[1,2]

The early studies relied on data from public sources and was known as 'Resource and Environmental Profile Analysis' (REPA) in the United States and 'Ecobalance' in Europe. Between 1970 and 1975, 15 REPAs were conducted and a protocol for performing the studies was developed. This methodology was reviewed by the US EPA and industrial experts.[2]

As a result of the oil crisis of the time, the early work focused on energy use rather than waste. By the early 1980s, however, environmental concerns gained prominence[1,2] and by the early 1990s, LCA had moved beyond inventory analysis towards impact assessment.[2] Since then, the Society of Environmental Toxicology and Chemistry (SETAC) has published articles relating to the technical framework of LCA, as well as its principles and methodology.[2–5]

Between 1997 and 2002, ISO developed standards for LCA resulting in the ISO 14000 series.

6.3 Basic principles of life cycle assessment

The basic concept of LCA relates to the assessment of environmental impacts of a product or service during its entire life span. ISO defines LCA as the 'compilation and evaluation of the inputs, outputs and the potential environmental impacts of a product system throughout its life cycle', and LCA studies analyse the environmental aspects and potential impacts throughout a product's life cycle (e.g. cradle to grave) from raw material acquisition through production, use and disposal.[6] A similar definition is employed in the United States.[7]

The basic concepts to consider for LCA are raw material extraction, manufacturing processes of product/s, transportation, distribution, retailing, use and disposal, as illustrated in Fig. 6.1. A typical LCA includes the different inputs involved in a product's life cycle system, such as the raw materials, chemicals, energy, water and other auxiliaries, as well as various outputs of the system, namely, the product; co-product/s; emissions to air, water and soil; waste; and other environmental releases.

LCA comprises four major phases, as defined by ISO, which are[6]:

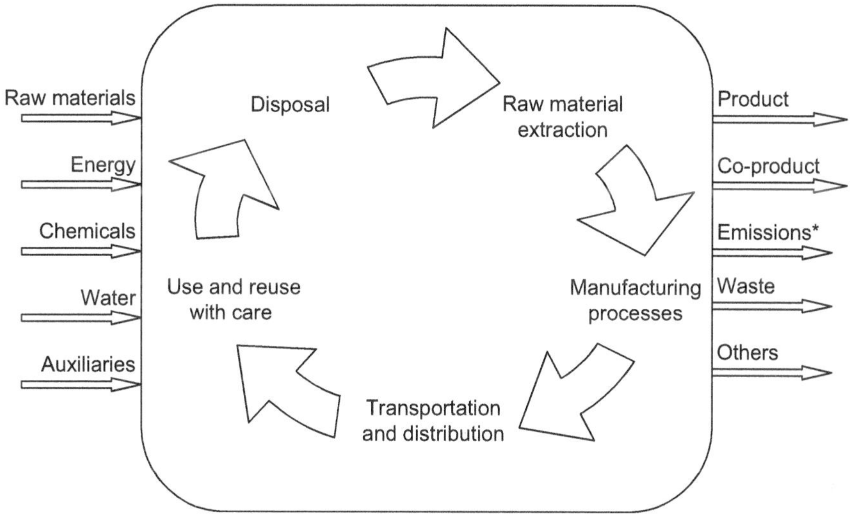

Figure 6.1 Basic concept of LCA. *Emissions include air, water and land emissions.

1. goal and scope definition,
2. LCI analysis,
3. life cycle impact assessment (LCIA),
4. life cycle interpretation.

The relations between these stages are depicted in Fig. 6.2, which shows LCA as an iterative process.

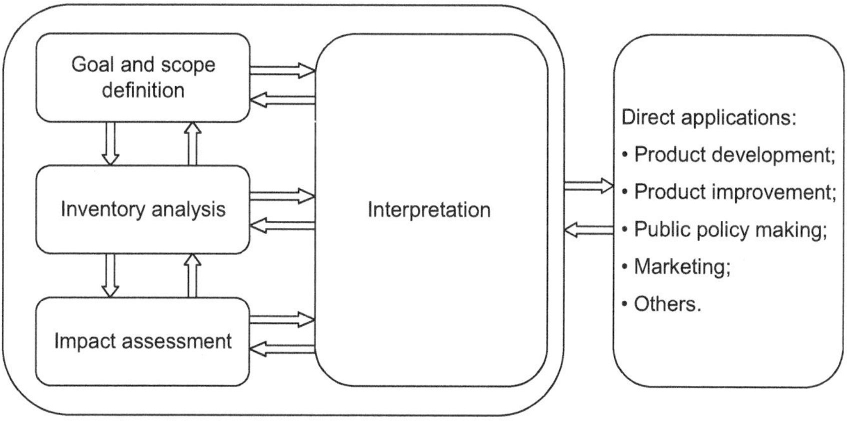

Figure 6.2 The general methodological framework of LCA, as per ISO 14040.6.

6.4 Life cycle assessment goal and scope definition

A typical LCA study begins with the definition of the goal and scope[8] and addresses the following questions:

- Why is the study being conducted? (intended applications)
- Who will the LCA results be communicated to? (intended audience)
- Whether the results are to be disclosed to the public?[8]

LCA can be conducted for a single product or can be used to compare different products/product groups. If it is a single product study, the goal may be to identify opportunities for improvement by identifying the hot spots in the life cycle. In the case of a comparative study, the goal may in addition be to identify the product with the lowest environmental impact.

Both types of study may be conducted for internal use only or may be used to communicate relevant information to the public. In the latter case, a critical review of the study by a panel of experts is mandatory for quality assurance purposes.

The definition of scope includes details of the type of product/s to be studied and the intended function/s of those product/s. Also considered at this stage are the following[6]:

- definition of the boundaries of the product/s system,
- stages of the life cycle to be included in the assessment (cradle to gate/cradle to grave/cradle to cradle/gate to gate),
- the functional unit of the study,
- categories of impact chosen for this study,
- allocation procedures followed in the study,
- cut-off criteria,
- type of LCIA methodology to be adopted,
- requirements of data and data quality,
- sources of data (primary and secondary sources to be clearly marked as such),
- uncertainty analysis,
- assumptions employed,
- type of LCA software/calculation methodology employed,
- limitations of the study,
- value choices and optional elements to be included,
- requirement for and type of critical review to be conducted,
- interpretation,
- type and format of the final report.

System boundaries should be clearly stated in the report and the justification or criteria for such a selection discussed.[9]

A functional unit, as per ISO 14040, is defined as the 'quantified performance of a product system for use as a reference unit'. Another term related to the functional unit is the reference flow, which is again defined by ISO as the, 'measure of the outputs from processes in a given product system required to fulfill the function expressed by the functional unit'.[6] For comparative studies, functional units must be consistent.[10]

Just to provide a textile example, a full LCA study cannot have a defined functional unit as one unit of T-shirt. This is not a complete functional unit. An example of a functional unit relating to a T-shirt case might be, 'A T-shirt made out of 100% cotton, to be worn 30 times in a one year life span'. This functional unit should further be supplemented with details of where the production of the T-shirt takes place, where it will be used and disposed of and how many times it will be washed and dried. Also included should be the type of washing and drying method used, temperature employed, washing chemicals used, etc.

Another critical parameter is allocation, defined by ISO 14040 as 'partitioning the input and output flows of a process or a product system between the product systems under study and one or more other product systems'.[6] An LCA practitioner should avoid allocation as per the guidelines of ISO.[9] In case it is not feasible to avoid allocation, methods recommended by ISO standards for LCA[11] should be employed. Allocation is a common problem in any product system and especially in the case of textile products. For instance, considering the cotton system, along with the cotton fibre, there are cotton seeds. Thus to model the LCA of cotton, the impacts to both fibre and seed should be allocated. In this case, allocation by mass or economics or a combination of both might be used. Allocation should be accounted for across all stages of the life cycle.

Another important factor to be specified is the cut-off criteria, defined by ISO as the 'specification of the amount of material or energy flow or the level of environmental significance associated with unit processes or product system to be excluded from a study'.[9] In practice, several cut-off criteria based on mass, energy and environmental impact can be followed in order to determine which inputs need to be included in the assessment and which outputs need to be traced to the environment.[9]

6.5 Life cycle inventory analysis

LCI analysis is defined by ISO as the 'phase of life cycle assessment involving the compilation and quantification of inputs and outputs for a product throughout its life cycle'.[6] The inventory relates to the compilation of various environmental inputs and outputs involved in the life cycle of a product. LCI analysis requires quantification of the following elements[2]:

- energy requirements,
- raw material needs,
- atmospheric emissions,
- waterborne emissions,
- emissions to land,
- solid wastes,
- other releases to the environment.

Inventory analysis translates in practice to data collection and analysis. Data collection involves the recording of the relevant inputs and outputs of the life cycle of a product or process.

The unit process is defined by ISO as the smallest element considered in the LCI analysis.[9] In the case of data collected from secondary sources such as technical documents or databases or government or scientific publications, the sources should be clearly referenced in the report.

The following are the four steps to be followed in an LCI analysis[2]:

1. Development of a flow diagram of the processes within the defined system boundary.
2. Development of a data collection methodology.
3. Collection of the relevant data.
4. Evaluation and reporting of results.

A flow diagram is a simple tool to aid in the mapping of the inputs and outputs to a process or system. Interrelationships between individual unit processes should be illustrated in order to build a picture of the life cycle in terms of the essential inputs and outputs.[2]Fig. 6.3 indicates a typical process flow diagram with generalized unit processes, and Fig. 6.4 depicts the LCI of a textile example (relating to cotton fibre manufacture).

Development of a data collection methodology begins with the design of a suitable data sheet to meet the goals and scope defined for the study. Data is then collected and validated, after which they should be related to the unit process and the functional unit assumed for the study.

Relevant data sources include, but are not limited to, test results from laboratories, metre readings from equipment, industry data sheets, reports, utility bills, journal papers, books, patents, previous LCI studies, government documents and equipment and process specifications.[2]

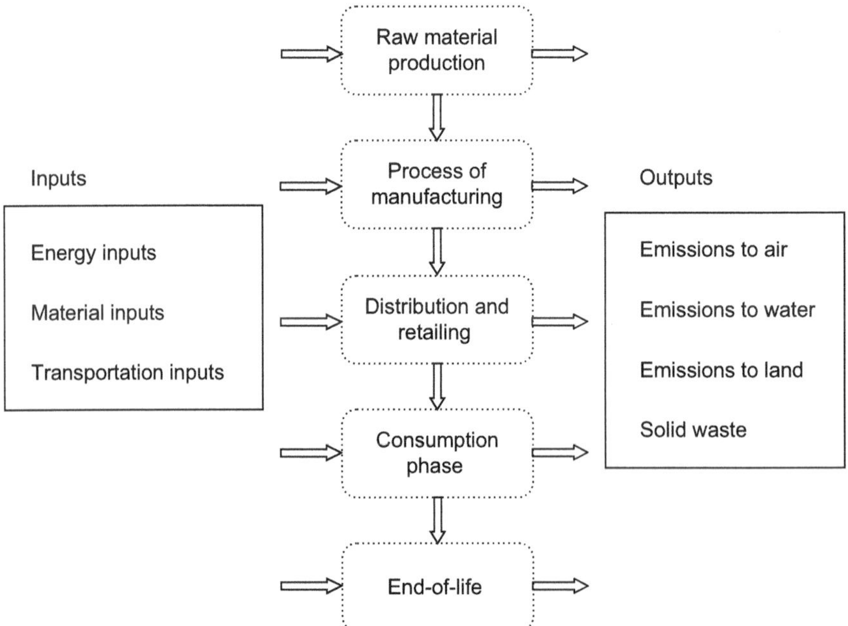

Figure 6.3 Generalized unit process flow diagram.

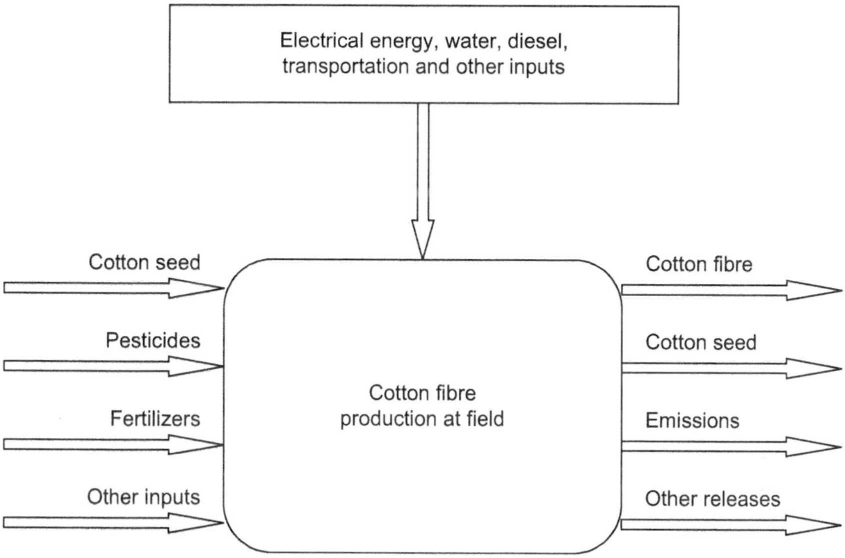

Figure 6.4 Life cycle inventory of cotton fibre.

6.6 Life cycle impact assessment

LCIA is defined by ISO as the phase of the process aimed at understanding and evaluating the magnitude and significance of the potential environmental impacts for a product system throughout the life cycle of that product.[9] This is the step that details the effects of resources consumed and emissions released into the atmosphere as identified in the LCI analysis. Effects on the environment and human health due to the activities (inputs and outputs) identified in the previous step are quantified in the LCIA.

For example, LCIA quantifies the impacts of pesticides and nutrients for the cultivation of cotton fibre, as depicted in Fig. 6.4. The following steps should be followed to conduct an LCIA[2]:

- *Selection and definition of impact categories* – selection of relevant environmental impact categories such as global warming or human toxicity.
- *Classification* – assignment of LCI results to the relevant impact categories. For example, classifying carbon dioxide emissions as relating to the global warming impact category.
- *Characterization* – quantitative modelling of life cycle impacts associated with emissions/resources. For example, modelling the potential impact arising from greenhouse gas emissions such as carbon dioxide and methane on the global warming impact category.
- *Normalization* – adjusting the expression of the characterized impact results to a common reference, enabling comparison of each factor.
- *Grouping* – relating different impact indicators in terms of factors such as location, local, regional or global.
- *Weighting* – cumulative expression of the results as a single environmental indicator. Weighing intends to emphasize the most important potential impacts amongst various impact categories studied.

- *Evaluating and reporting LCIA results* – evaluation of the LCIA results to achieve a better understanding of their reliability.[2]

Selection of impact categories, classification and characterization are mandatory elements of an LCIA as per ISO 1404X standards. Normalization, grouping and weighing are optional elements in an LCIA under the same standard.

6.6.1 Mandatory elements of life cycle impact assessment: definition and classification of impact categories

In LCIA, there are numerous impact categories including depletion of abiotic resources, climate change, human toxicity, acidification, eutrophication, ecotoxicity, photo-oxidant formation, stratospheric ozone depletion, land use, water depletion, depletion of minerals and use of fossil fuels.

A schematic cause pathway needs to be developed for each substance emerging from the LCI analysis that describes the environmental mechanism of impact for the emitted substance.[12,13] Two different approaches exist: the so-called midpoint and endpoint approaches.[12]

The LCIA midpoint approach is also known as the problem-oriented approach or classical impact assessment method.[14–17] It translates category impacts into real phenomenon-based environmental themes such as climate change, acidification and human toxicity.[18] The endpoint approach is also known as the damage-oriented approach.[17] It translates environmental impacts into major issues of concern such as human health, ecosystem quality (quality of the natural environment), natural resources and human impact.[12,19] A number of methods are available to quantify life cycle impacts. These include the Eco-indicator'99 method, EPS 2000 and the Recipe endpoint approach, which are examples of endpoint assessment methodologies. The EDIP 2003, CML 2001, TRACI and the Recipe midpoint approach are some of the commonly used midpoint assessment methodologies. Fig. 6.5 depicts the difference between midpoint and endpoint assessment methodologies.

6.6.2 Mandatory elements of life cycle impact assessment: characterization

The final step in LCIA is characterization, followed by the selection and definition of relevant impact categories. In this step, the LCI results will be converted into common units and aggregated within each impact category. The conversion process utilizes characterization factors.[9] Characterization is calculated according to[21]

$$IR_c = \Sigma\, CF_{cs} * m_s \qquad [6.1]$$

where IR_c is the indicator result for the impact category, CF_{cs} is the characterization factor that connects intervention s with impact category c and m_s is the size of intervention (i.e. the mass of substance emitted). For example, the characterization equation for calculating climate change in terms of global warming potential is given by[22]

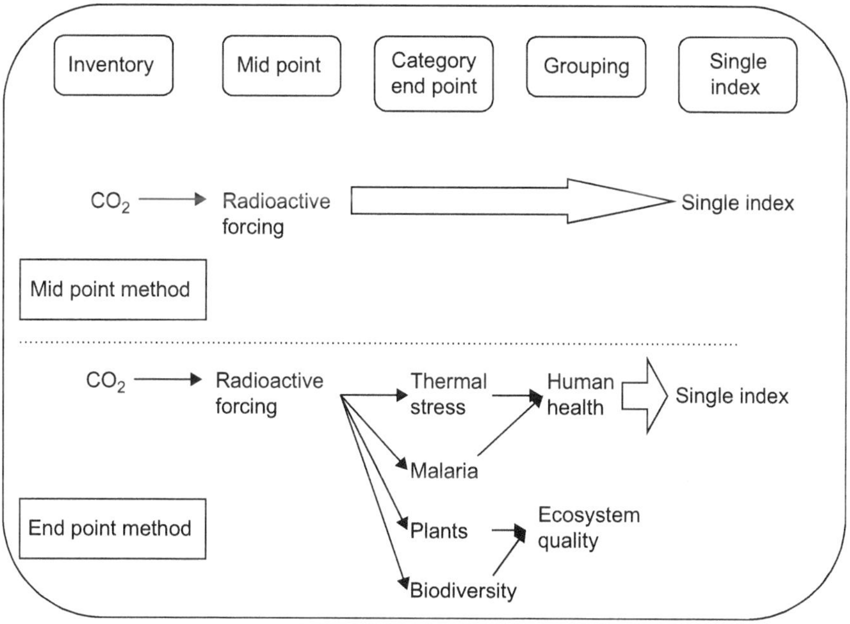

Figure 6.5 Differences between midpoint and endpoint assessment methodologies.[20]

$$\text{Climate change} = \Sigma\ GWP_{a,i} * m_i \qquad [6.2]$$

The indicator result is expressed in kilograms of the reference substance, in this case CO_2. $GWP_{a,i}$ is the global warming potential for substance *i* integrated over a specified number of years, while m_i (kg) is the quantity of substance *i* emitted.[22]

6.6.3 Optional elements of life cycle impact assessment: normalization

An optional step of LCIA as per the ISO standards is normalization. This step characterizes impact indicator results expressed relative to some relevant baseline.[9] This process aids in understanding the relative magnitude of each indicator result and places the characterized impact indicator results into a broader context.[9,23] Examples of reference values include the total inputs and outputs for a given area (at a global, regional, national or local level), the total inputs and outputs for a given area on a per capita basis and the inputs and outputs in an alternative product system.[9] The normalization factors are chosen to portray the magnitude of an impact category for a geographical location and over a certain period. Normalization proceeds as follows:

$$N_c = S_c/R_c \qquad [6.3]$$

where c represents the impact category, N is the normalized indicator, S is the characterized indicator obtained from the characterization step and R is the normalization factor or reference value.

6.6.4 Optional elements of life cycle impact assessment: grouping

Another optional step is the sorting and/or ranking of impact categories. This involves assignment of the impact categories into one or more sets. Impact categories can be sorted based either on the characteristics of the inputs and outputs or by global/local/regional scales or according to a given hierarchy, such as in terms of high, medium or low priority.[9]

6.6.5 Optional elements of life cycle impact assessment: weighing

Weighing converts the normalized impact indicator results by using weighing factors. Weighing factors are value choices and are dependent on subjective evaluations.[9,23] There are various methods for weighing, such as the proxy approach, distance to target method, monetarization method and panel approach method. Details can be found in Refs. 24–27.

6.7 Life cycle interpretation

The last step in the LCA process is the interpretation of results. There are three important elements that need to be considered here[9]:

1. identification of significant issues;
2. evaluation by sensitivity, consistency and completeness check;
3. conclusion, limitations and recommendations.

6.8 Standards for life cycle assessment

The ISO 14040 series for LCA are listed in Table 6.1.

6.9 Different life cycle assessment methods

6.9.1 Attributional and consequential life cycle assessment

There are two main approaches to conduct an LCA, the consequential and attributional methods. Attributional LCA (ALCA) requires description of the environmental

Table 6.1 ISO standards for life cycle assessment[8]

Number	Title	Year of introduction	Current status
14040	Principles and framework	1996	Updated in 2006. Current version is introduced in 2006.
14041	Goal and scope definition and inventory analysis	1998	Updated in 2006 and merged into 14044.
14042	Life cycle impact assessment	2000	Updated in 2006 and merged into 14044.
14043	Life cycle interpretations	2000	Updated in 2006 and merged into 14044.
14044	Requirements and guidelines	2006	Replaces 14041, 14042 and 14043.
14047	Examples of application of ISO 14042	2003	
14048	Data documentation format	2001	
14049	Examples of application of ISO 14041	2000	

attributes of a life cycle (environmentally relevant physical flows to and from a life cycle) and its sub-systems. Consequently, ALCA concentrates on describing the effects of changes within the life cycle and the impact of such according to a set of possible decisions.[28,29]

ALCA is also known as the 'business-as-usual' scenario, as the values used are average values based on current trends and normal business practices. Consequential LCA (CLCA) focuses, however, on predicting the consequences if changes are made to the established processes. It addresses the indirect effects of these changes, along with the direct effects.[30]

From the ALCA, an understanding can be gained of the impacts of the processes used to produce, consume and dispose of a product without considering the indirect effects arising from changes in the level of output of the product under question. Details of the average unit of product can be obtained from ALCA, and hence, it is useful for practices such as consumption-based carbon accounting. ALCA studies rely on the stoichiometric relationships between inputs and outputs; hence, the results derived from ALCA have known accuracy. It is possible to identify opportunities to reduce the direct impacts arising from the various life cycle stages of a product.[31]

With CLCA, an understanding of the effects of changes in the level of output (including consumption and disposal) of a product can be gained. CLCA studies are dependent on economic models (dealing with the relationships amongst demand for inputs, elasticity of prices, supply and market effects of co-products); hence, the results derived from CLCA studies need to be used with caution because the accuracy of the results are not well known.[31] Attributes of LCA models pertaining to ALCA and CLCA and the differences in results due to these two different approaches are explained in the study by Thomassen *et al.*[32]

6.9.2 Screening and detailed life cycle assessment

In terms of level of detail, there are two quantitative types of LCA: screening and full or detailed LCA.

Screening LCA is conducted when quick results are required and the aim is to understand where the major environmental impacts of a product/process/service are across the life cycle. This kind of study can be performed in the absence of primary data and may rely entirely on secondary data, and as such is usually used for internal purposes only.

Full or detailed LCA, as the name implies, goes through all the phases of an LCA and adheres strictly to the ISO 1404X standards. A full LCA study is based on primary data, with the aid of secondary data, as stipulated in the ISO standards, and may be communicated to the public.

6.10 Different life cycle impact assessment methods

In any LCA study, LCIA plays an important role and there are many methods available to quantify various life cycle impact indicators. Each method is unique in terms of its calculation methodology. Some of the most common methods are

- Eco-indicator'99
- CML 2001
- EDIP 2003
- EPS 2000
- EPD 2007
- Ecological Scarcity 2006
- Impact 2002+
- Recipe
- TRACI
- Ecological Scarcity Method
- Single indicator methods such as ecological and carbon footprints
- ILCD 2011
- USEtox

6.10.1 Eco-indicator'99 method

As seen earlier, LCIA has many mandatory and optional elements as per the ISO standards for LCA, one of which is the single score analysis. This allows expression of the environmental load of a product/process/service via a single score, which may be used

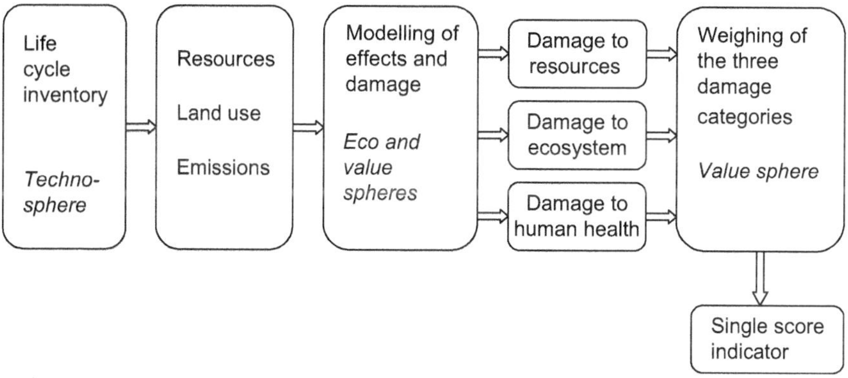

Figure 6.6 Core principle of Eco-indicator'99.

for internal purposes. Eco-indicator'99 is one of the most widely used and well-known methods of LCIA.

The details of Eco-indicator'99 can be found in Refs. 18, 33–35. The approach essentially calculates environmental load in terms of damages (endpoints) relating to three main categories:

- human health,
- ecosystem quality,
- resources.

The three damage models developed and used in Eco-indicator'99 are listed in the following:

- Damage to human health expressed as disability-adjusted life-years (DALY). DALY is an index used by WHO and the World Bank and is derived from a combination of the number of years lost and the number of years living in a disabled condition.[36,37]
- Damage to the ecosystem quality is expressed as the percentage of species that have become extinct in a given area during a certain period due to environmental load.[36,37]
- Damage to resources is expressed in terms of the surplus energy needed for future extractions of minerals and fossil fuels.[36]

The core concept of the Eco-indicator'99 method is illustrated in Fig. 6.6.

As shown in Fig. 6.6, the core principles of the Eco-indicator method are the techno, eco and value spheres.[35] Description of life cycle phases, emissions accounting from the inventory and dealing with allocation issues are dealt with in the techno sphere and the damages are modelled in the eco sphere. The impacts and effects of changes take place in the value sphere.[35,37]

6.10.1.1 Characterization

Different impact indicators are characterized in this approach and each impact indicator belongs to one of the three damage categories defined in this modelling system. The following are the impact categories dealt with by the Eco-indicator'99 methodology.[36]

6.10.1.1.1 Emissions

- *Carcinogens* – emissions of carcinogenic substances to air, water and soil. Damage is expressed in DALY/kg emission.
- *Respiratory organics* – respiratory effects arise from summer smog, which is mainly due to the emissions of organic *substances* into the air. Damage is expressed in DALY/kg emission.
- *Respiratory inorganics* – respiratory effects mainly due to winter smog, which is caused by the emission of dust, sulphur and nitrogen oxides into the air. Damage is expressed in DALY/kg emission.
- *Climate change* – impacts are expressed in DALY/kg emission caused by the effects of climate change.
- *Radiation* – impacts of radiation, expressed in DALY/kg emission.
- *Ozone layer depletion* – due to increased UV radiation resulting from the emission of ozone-depleting substances into the air and expressed in terms of DALY/kg emission.
- *Ecotoxicity* – damage to ecosystem quality, expressed in Potentially Affected Fraction (PAF)*m^2*year/kg emission. This is mainly due to the emission of ecotoxic substances into air, water and soil.
- *Acidification/eutrophication* – damage to the ecosystem expressed in Potentially Disappeared Fraction (PDF)*m^2*year/kg emission and arising from the emission of acidifying substances into the air.

6.10.1.1.2 Land use

Damage due to either conversion or occupation of land or occupation, expressed in PDF*m^2*year/m^2 or m^2a.[36]

6.10.1.1.3 Resource depletion

Resources become harder to extract and this category is defined by the 'surplus energy'. Surplus energy is quantified for minerals and fossil fuels:

- *Minerals* – surplus energy per kilogram of mineral or ore due to decreasing ore grades.
- *Fossil fuels* – surplus energy per extracted megajoule, kilogram or cubic metre of fossil fuel, as a result of lower quality resources.[36]

Table 6.2 lists the individual impacts grouped under different damage categories selected in this methodology.

6.10.1.2 Dealing with uncertainties

Two types of uncertainties are associated with the calculation of indicators using the Eco-indicator'99 methodology. These are related either to the models used or to the quality of the data collected. Data-related uncertainties are usually expressed as the squared geometric standard deviation in the original results. Uncertainties pertaining to the model are due to subjective choices employed in the model.[36–38] Three methodological approaches have been developed to deal with those uncertainties:

- The egalitarian perspective, in which damages cannot be avoided and the model considers the long timescale effects. It is assumed in this perspective, for example, that fossil fuels cannot be substituted. In the DALY calculations, age weighing is not included.

Table 6.2 Impact indicators under damage categories – Eco-indicator'99 method

Damage to human health	Damage to ecosystem quality	Damage to resources
Carcinogens	Land use	Minerals extraction effects
Respiratory organics	Acidification/eutrophication	Fossil fuel extraction effects
Respiratory inorganics	Ecotoxicity	
Climate change		
Radiation		
Ozone layer depletion		

- The hierarchist perspective, in which damages are assumed to be avoidable by good management and the timescale considered is again long. In this case, fossil fuels are assumed not to be easily substituted. Age weighing is not included in the DALY calculations.
- The individualist perspective, in which damages are assumed to be avoided by technological advancements and economic development. This perspective considers a short timescale of 100 years or less. In this case, fossil fuels are assumed not to be exhausted and hence are not included in the methodology. Age weighing is included in the DALY calculations.[36–38]

6.10.1.3 Normalization and weighing

Normalization and weighing are performed at a damage category level (i.e. endpoint level) with the data for normalization calculated based on an average European level with 1993 as the base year. Weighing is performed for three damage categories with a specific weighing to each individual category.[36–38]

6.10.2 Institute of Environmental Sciences (CML), Leiden University, 2001 method

In 2001, the Institute of Environmental Sciences (CML) of Leiden University developed a set of impact categories for the midpoint approach together with characterization methods for LCIA.

6.10.2.1 Characterization

CML categorizes the impact categories according to three types:

1. Obligatory impact categories, which are included in most LCA studies.
2. Additional impact categories, for which operational indicators exist and are not often included in LCA studies.
3. Other impact categories, for which no operational indicators exist and are therefore impossible to include in LCA studies.[36,39]

The following impact categories are of the obligatory type and a short description is included for each in Table 6.3. Details can be found in Refs. 36 and 39.

Apart from the above-mentioned, the following impact categories are also available:

- Global warming (measured over different time frames).
- Upper limit for net global warming.
- Lower limit for net global warming.
- Ozone layer depletion (measured over different time frames).
- Human toxicity (measured over different time frames).
- Freshwater aquatic ecotoxicity (measured over different time frames).
- Marine aquatic ecotoxicity (measured over different time frames).
- Terrestrial ecotoxicity (measured over different time frames).
- Marine sediment ecotoxicity (measured over different time frames).
- Average European (kg NO_x-eq), average European (kg SO_2-eq).
- Land competition.
- Ionizing radiation.
- Photochemical oxidation, photochemical oxidation (low NO_x).
- Malodorous air.
- Equal benefit incremental reactivity.
- Maximum incremental reactivity, maximum ozone incremental reactivity.[37]

6.10.2.2 Normalization and weighing

In the CML 2001 method, normalization scores are calculated for three reference situations and for each baseline indicator. The indicators are the World in 1990, Europe in 1995 and the Netherlands in 1997.[37]

6.10.3 The Recipe method

The Recipe approach is the most recent impact assessment method for LCIA. It was developed with 18 midpoint indicators and 13 endpoint indicators.[40,41] Midpoint indicators dealt with by the Recipe method are as follows:

1. Climate change (CC)
2. Ozone depletion (OD)
3. Terrestrial acidification (TA)
4. Freshwater eutrophication (FE)
5. Marine eutrophication (ME)
6. Human toxicity (HT)
7. Photochemical oxidant formation (POF)
8. Particulate matter formation (PMF)
9. Terrestrial ecotoxicity (TET)
10. Freshwater ecotoxicity (FET)
11. Marine ecotoxicity (MET)
12. Ionizing radiation (IR)
13. Agricultural land occupation (ALO)
14. Urban land occupation (ULO)

Table 6.3 Obligatory impact categories of the CML 2001 method[36,39]

Environmental indicators	Description	Unit	Geographical scope of the indicator
Depletion of abiotic resources	Measurement of depletion of non-renewable resources (extraction of minerals and fossil fuels)	kg antimony equivalents/kg extraction	Global scale
Climate change	Measurement of greenhouse gas emissions to air (such as CO_2 and methane) that cause global warming	kg equivalent CO_2/kg extraction	Global scale
Stratospheric ozone depletion	Measurement of the gases that potentially deplete ozone layer	kg CFC-11 equivalent/kg emission	Global scale
Human toxicity	Measurement of the potential toxicity of materials based on the chemical condition, original emission place and its fate	1,4-Dichlorobenzene equivalents/kg emission	Between local and global scales depending on the fate of the substance
Freshwater aquatic ecotoxicity	Measurement of the potential toxicity of materials based on the chemical condition, original emission place and its fate	1,4-Dichlorobenzene equivalents/kg emission	Global/continental/regional and local scale
Marine aquatic ecotoxicity	Measurement of the potential toxicity of materials based on the chemical condition, original emission place and its fate	1,4-Dichlorobenzene equivalents/kg emission	Global/continental/regional and local scale
Terrestrial aquatic ecotoxicity	Measurement of the potential toxicity of materials based on the chemical condition, original emission place and its fate	1,4-Dichlorobenzene equivalents/kg emission	Global/continental/regional and local scale
Photo−oxidant formation	Measurement of the formation of reactive substances (mainly ozone) that impact human health, ecosystems and crops	kg ethylene equivalents/kg emission	Between local and continental scale
Acidification	Measurement of emissions that cause acidifying effects to the environment	kg SO_2 equivalents/kg emission	Between local and continental scale
Eutrophication	Measure of emissions that cause eutrophying effects to the environment (fertilizers from fields)	kg PO_4 equivalents/kg emission	Between local and continental scale

15. Natural land transformation (NLT)
16. Water depletion (WD)
17. Mineral resource depletion (MRD)
18. Fossil fuel depletion (FD).[42]

The following are the main endpoint indicators used by the Recipe method:

- Damage to human health (HH)
- Damage to ecosystem diversity (ED)
- Damage to resource availability (RA).[42]

6.10.4 *Single indicator methods such as ecological and carbon footprints*

LCIA may focus on quantifying a single indicator such as cumulative energy demand, the ecological footprint, the carbon footprint measured by the IPCC 2007 method over 20-, 100- and 500-year time horizons, and ecosystem damage potential.

6.10.5 *International Reference Life Cycle Data System method*

ILCD stands for International Reference Life Cycle Data System and it results from a project conducted by the Joint Research Centre (JRC) of the European Commission, which analysed several LCIA methodologies to reach consensus on the recommended method for each environmental theme, at both midpoint and endpoints.[43]

Many impact category groups mentioned in the following are included under this method at both midpoint and endpoints, and these intermediate endpoint categories have been grouped into three areas of protection, namely, human health, resource depletion and ecosystems[43]:

- ✓ Acidification
- ✓ Climate Change
- ✓ Depletion of Abiotic Resources
- ✓ Ecotoxicity
- ✓ Eutrophication
- ✓ Human Toxicity
- ✓ Ionizing Radiation
- ✓ Land Use
- ✓ Ozone Depletion
- ✓ Particulate Matter
- ✓ Photochemical Oxidation

6.10.6 *USEtox*

USEtox is one of the recent methods that is based on scientific consensus to calculate environmental impact to identify and obtain human and eco-toxicological impact values of chemicals in LCIA. The main output includes a database of recommended and interim characterization factors including environmental fate, exposure and effect parameters for human toxicity and ecotoxicity.[43]

The impact categories dealt by this method are

- ecotoxicity
- human toxicity

6.11 Tools for the calculation of life cycle assessment studies

There are a number of software tools currently available to model the life cycle impacts of any product/process/service. Some of the most popular and widely known tools used for LCA are:

- SimaPro
- GaBi
- Umberto
- Eco-it
- TEAM
- KCL-ECO
- The Boustead Model
- WISARD
- Open-LCA.

A comprehensive list of tools for LCA can be found in the LCA resources directory of the European Commission's website.[44]

6.12 Advantages and limitations of life cycle assessment

LCA provides information on the multiple environmental impacts created by a product from its cradle to grave. A product may be declared as a green product if it is evaluated to be so by LCA.

Some of the main advantages of LCA are listed in the following:

- LCA can systematically analyse and address the environmental profile of a product from its raw material extraction stage through to the disposal stage using a holistic approach.
- LCA can be used to compare different products and product systems, as well as alternative processes, to identify the best choice in terms of environmental impacts.
- LCA is useful in quantifying the extraction of resources and emissions of a product system or process to air, water and land and their associated impacts.
- LCA can be used to identify hot spots in the life cycle of a product or for different production processes of a product system.
- Based on the information supplied by LCA, LCIA and hot-spot analysis, it is viable for a manufacturer to reduce the environmental impacts of a product system or production processes.
- It is a comprehensive tool as far as environmental interventions of a product system or a process is concerned.[45]

- LCA is a viable tool to select the best waste management approach amongst different options.[46]
- LCA can offer clear business benefits such as[46]
  - lower costs and increasing revenues,
  - brand enhancement and corporate,
  - better market access,
  - quantitative basis for corporate sustainability,
  - enhanced customer relationships.
- LCA can be used as a tool for companies to evaluate their position in terms of sustainability, and based on LCA results, they can define their sustainability goals.
- LCA can be used to raise environmental awareness amongst employees, customers and the public.
- LCA studies are useful for public communication (Eco-labelling of products, Environmental Product Declarations, etc.).
- LCA studies can be used in companies to establish performance and to manage risks within the supply chain.
- LCA studies can be used for environmental reporting, green procurement and regulatory compliance.
- LCA studies can be employed for industrial benchmarking.

Despite these potential benefits, LCA studies are subject to criticism including the following:

- An LCA study is very complex and highly data intensive.[45]
- LCA studies are performed using different assumptions and subjective valuation procedures.[45]
- LCA studies address only potential impacts, not actual impacts, and require a high degree of expertise.[45]
- The availability of quality data for LCA is problematic.
- LCA studies address only environmental aspects of a product and not economic or social aspects of a product.[47]
- LCA cannot address localized impacts and it is a steady-state approach rather than a dynamic one.[47]
- LCA models focus principally on industrial activities and do not consider market mechanisms or secondary effects on technological development.[47]
- Results derived from a detailed LCA study are location specific and are not directly applicable to other locations.
- Availability of relevant and up-to-date databases is a major issue.
- Transparency may be lacking or perceived to be lacking.
- LCA studies suffer uncertainties such as data uncertainty and uncertainty of models.
- LCA has restrictions in its applicability as a decision-support tool in planning waste management and policy making.[48]

6.13 Summary

LCA is a methodological tool employed to analyse the environmental impacts of a product from its cradle to grave and can be applied to a process or a service system. In this chapter, the principles and concepts of LCA have been discussed through its four stages: goal and scope definition, LCI, LCIA and life cycle interpretation. Different standards for LCA, various methods of conducting LCA studies and models used for LCA have been detailed.

6.14 Sources of further information and advice

- Arnold FS. Why environmental life cycle assessment doesn't work. *Journal of Environmental Law and Practice*. 1995;2(5):4–14.
- Assies JA. Introduction paper to SETAC-Europe workshop on environmental life cycle analysis of products. In: *Life-Cycle Assessment, Proceedings of a SETAC-Europe Workshop on Environmental Life Cycle Assessment of Products, 2–3 December 1991, Leiden*. Brussels, Belgium: SETAC-Europe.
- Ayres RU. Life cycle analysis: a critique. *Resources, Conservation and Recycling*. 1995;14: 199–223.
- Bare JC, Pennington DW, Udo de Haes HA. Life cycle impact assessment sophistication, international workshop. *International Journal of Life Cycle Assessment*. 1999;4(5): 299–306.
- Baumann H, Tillman AM. *A Hitchhiker's Guide to Life Cycle Assessment*. Lund: Studentlitteratur; 2004.
- Berkhout F, Howes, R. The adoption of life-cycle approaches by industry: patterns and impacts. *Resources, Conservation and Recycling*. 1997;20:71–94.
- Bilitewski B, Winkler J. Comparative evaluation of life cycle assessment models for solid waste management. *Waste Management*. 2007;27(8):1021–1031.
- Boustead I. LCA – How it Came About, The Beginning in the UK. *International Journal of Life Cycle Assessment*. 1996;1(3).
- Clift R, Doig A, Finnveden, G. The application of life cycle assessment to integrated solid waste management, Part I – methodology. *Transactions IchemE*. 2000;78 (part B): 279–287.
- Curran MA. *Environmental Life-Cycle Assessment*. New York: McGraw-Hill; 1996.
- Denison RA. Environmental life-cycle comparisons of recycling, landfilling and incineration. A review of recent studies. *Annual Review of Energy and the Environment*. 1996;21: 191–237.
- Ekvall T, Finnveden G. The application of life cycle assessment to integrated solid waste management. Part 2: Perspectives on energy and material recovery from paper. *Transactions of the Institution of Chemical Engineers*. 2000;78(4):288–294.
- Ekvall T, Finnveden G. Allocation in ISO 14041 – a critical review. *Journal of Cleaner Production*. 2001;9:197–208.
- Ekvall T. Cleaner production tools: LCA and beyond. *Journal of Cleaner Production*. 2002; 10:403–406.
- Ekvall T, Ciroth A, Hofstetter P, Norris G. Evaluation of attributional and consequential life cycle assessment. In: *Working Paper Distributed at 14th SETAC-Europe Annual Meeting, 18–22 April 2004*; 2004.
- Ekvall T, Andrae A. Attributional and consequential environmental assessment of the shift to lead-free solders. *International Journal of Life Cycle Assessment*. 2006;11(5):344–353.
- Finnveden G. On the limitations of life cycle assessment and environmental systems analysis tools in general. *International Journal of Life Cycle Assessment*. 2000;5:229–238.

- Finnveden G, Hauschild MZ, Ekvall T, Guinee J, Heijungs R, *et al.* Recent developments in life cycle assessment. *Journal of Environmental Management.* 2009;91(1):1–21.
- Frankl P, Rubik F. *Life Cycle Assessment in Industry and Business.* Berlin: Springer-Verlag; 2000.
- Frankl P, Rubik F, editors. *Life Cycle Assessments in Industry and Business, Adoption Patterns, Applications and Implications.* Berlin, Heidelberg, New York: Springer; 2000.
- Guinee JB, Heijungs R, Huppes G, Zamagni A, Masoni R, *et al.* Life cycle assessment: past, present, and future. *Environmental Science and Technology.* 2011;45(1):90–96.
- Hauschild MZ. Assessing environmental impacts in a life cycle perspective. *Environmental Science and Technology.* 2005;39:81A–88A.
- Heijungs R, Guinee JB, Huppes G, Lankreijer RM, Udo de Haes HA, *et al. Environmental Life Cycle Assessment of Products. Guide and Backgrounds – October 1992.* Leiden, The Netherlands: Centre of Environmental Science, Leiden University; 1992.
- Heijungs R. Identification of key issues for further investigation in improving the reliability of life-cycle assessments. *Journal of Cleaner Production.* 1996;4:159–166.
- Heijungs R, Goedkoop M, Struijs J, Effting J, Sevenster M, Huppes G. *Towards a Life Cycle Impact Assessment Method Which Comprises Category Indicators at the Midpoint and the Endpoint Level.* Report of the first project phase: design of the new method; 2003.
- Heiskanen E. The institutional logic of life cycle thinking. *Journal of Cleaner Production.* 2002;10:427–437.
- Huijbregts MAJ. Application of uncertainty and variability in LCA. Part I: A general framework for the analysis of uncertainty and variability in life cycle assessment. *International Journal of Life Cycle Assessment.* 1998;3:273–280.
- Hunt R, Selelrs J, Franklin, W. Resource and environmental profile analysis: a life cycle environmental assessment for products and procedures. *Impact Assessment Review.* 1992; 12:245–269.
- Hunt RG, Franklin WE. LCA – how it came about – personal reflections on the origin and the development of LCA in the USA. *International Journal of Life Cycle Assessment.* 1996; 1(1):4–7.
- Hunt R, Franklin W. Pollute or not pollute, that is the question. *Pulp and Paper.* 1976; 60(11):80–82.
- Krozer J, Viz JC. How to get LCA in the right direction. *Journal of Cleaner Production.* 1998;6:53–61.
- Molina-Murillo SA, Smith TM. Exploring the use and impact of LCA-based information in corporate communications. *International Journal of Life Cycle Assessment.* 2009;14(2): 184–194.
- Prague RH, Guinée JB. *Environmental Life Cycle Assessment of Products, Guide (Part 1) and Backgrounds (Part 2).* Leiden: CML, TNO B&B; 1992.
- Product Life Cycle Assessment – Principles and Methodology; Nord 1992: 9, Nordic Council of Ministers: Copenhagen, Denmark, 1992.
- Quella F, Schmidt WP. Integrating environmental aspects into product design and development—the new ISO TR 14062. *International Journal of Life Cycle Assessment.* 2003;8:113–114.
- Schmidt WP, Sullivan J. Weighing in life cycle assessments in a global context. *International Journal of Life Cycle Assessment.* 2002;7(1):5–10.
- Udo de Haes HA, Jolliet O, Finnveden G, Hauschild M, Krewitt W, Mueller-Wenk R. Best available practice regarding impact categories and category indicators in life cycle assessment. *International Journal of Life Cycle Assessment.* 1999;4(2–3).

- UNEP Industry and Environment. *Towards the Global Use of Life Cycle Assessment.* Paris, France: United Nations Publication Sales no. 92-807-1740-5; 1999.
- Wenzel H, Hauschild M, Alting L. *Environmental Assessment of Products, Volume 1: Methodology, Tools and Case Studies in Product Development.* London: Chapman and Hall; 1997.
- Wrisberg N, de Haes H, Helias UA, editors. *Analytical Tools for Environmental Design and Management in a Systems Perspective.* Dordrecht: Kluwer Academic Publishers; 2002.

6.15 References

1. Jensen AA, et al. *Life Cycle Assessment (LCA) – A Guide to Approaches, Experiences and Information Sources. Environmental Issues Series, No. 6.* European Environment Agency; August 1997.
2. Curran MA. *Life Cycle Assessment: Principles and Practice.* Ohio, USA: US Environmental Protection Agency; 2006.
3. Fava J, Denison R, Jones B, et al., eds. *A Technical Framework for Life Cycle Assessment.* Society of Environmental Toxicology and Chemistry; 1991.
4. Consoli F, Allen D, Boustead I, Fava J, Franklin W, et al., eds. *Guidelines for Life Cycle Assessment: A 'Code of Practice'.* Society of Environmental Toxicology and Chemistry; 1993.
5. Barnthouse L, Fava J, Humphreys K, Hunt R, Laibson L, et al., eds. *Life Cycle Impact Assessment: The State-of-the-Art.* Society of Environmental Toxicology and Chemistry; 1997.
6. International Standards Organization. *Environmental Management – Life Cycle Assessment – Principles and Framework.* 2nd ed July 01, 2006. ISO 14040.
7. Environmental Protection Agency. *Life Cycle Assessment: Inventory Guidelines and Principles.* EPA/600/R-92/245. Cincinnati, Ohio, USA: Office of Research and Development; 1993.
8. Heijungs R, Guinee JB. An overview of the life cycle assessment method – past, present, and future. In: Curran MA, ed. *Life Cycle Assessment Handbook: A Guide for Environmentally Sustainable Products.* Scrivener Publishing LLC; 2012:15–42.
9. International Standards Organization. *Environmental Management – Life Cycle Assessment – Requirements and Guidelines.* ISO 14044. 1st ed July 01, 2006.
10. Rebitzer G, et al. Life cycle assessment. Part 1: framework, goal and scope definition, inventory analysis, and applications. *Environment International.* 2004;30:701–720.
11. Luo L, van der Voet E, Huppes G, Udo de Haes HA. Allocation issues in LCA methodology: a case study of corn stover-based fuel ethanol. *International Journal of Life Cycle Assessment.* 2009;14(6):529–539.
12. PE International. *Life Cycle Assessment – LCA Methodology.* Available at: http://www.pe-international.com/topics/life-cycle-assessment-lca-methodology/.
13. Bare JC, Hofstetter P, Pennington DW, Udo de Haes HA. Midpoints versus endpoints: the sacrifices and benefits. *International Journal of Life Cycle Assessment.* 2000;5(6):319–326.
14. Jolliet O, Mueller-Wenk RA, Peña C, et al. *Life Cycle Impact Assessment Programme of the Life Cycle Initiative*; 2003. Final report of the LCIA Definition Study [Electronic Version]. Obtained from: http://lcinitiative.unep.fr/includes/file.asp?site=lcinitandfile=F7BF1ABF-8B98-4A95-9FDE-3E32EB7C4EC4.
15. Ortiz O, Francesc C, Sonnemann G. Sustainability in the construction industry: a review of recent developments based on LCA. *Construction and Building Materials.* 2009;23:28–39.

16. Jolliet O, Müller-Wenk R, Bare J, et al. The LCIA midpoint-damage framework of the UNEP/SETAC life cycle initiative. *International Journal of Life Cycle Assessment*. 2004; 9(6):394–404.
17. Dreyer LC, Niemann AL, Hauschild MZ. Comparison of three different LCIA methods: EDIP97, CML2001 and Eco-indicator'99. Does it matter which one you choose? *International Journal of Life Cycle Assessment*. 2003;8(4):191–200.
18. Sleeswijk AW, Van Oersc LFCM, Guinée JB, Struijsd J, Huijbregts MAJ. Normalisation in product life cycle assessment: an LCA of the global and European economic systems in the year 2000. *The Science of the Total Environment*. 2008;390:227–240.
19. Bare JC, Gloria TP. Environmental impact assessment taxonomy providing comprehensive coverage of midpoints, endpoints, damages and areas of protection. *Journal of Cleaner Production*. 2008;(16):1021–1035.
20. Itsubo N, Inaba A. Assessment of environmental impact of manufacturing steel considering physical damage to human health. *Materials Transactions*. 2003;44(1):167–172.
21. Heijungs R, Koning A. *Improvement of LCA Characterization Factors and LCA Practice for Metals*. TNO Report- E 2004/347. International Council on Minings and Metals; August 2004. Obtained from: http://media.leidenuniv.nl/legacy/final%20report%20metals.pdf.
22. Guinée JB, Gorrée M, Heijungs R, Huppes G, Kleijn R, et al. *Handbook on Life Cycle Assessment. Operational Guide to the ISO Standards. I: LCA in Perspective. IIa: Guide. IIb: Operational Annex. III: Scientific Background*. Dordrecht: Kluwer Academic Publishers; 2002. ISBN 1-4020-0228-9, 692 pp.
23. Menoufi KAI. *Life Cycle Analysis and Life Cycle Impact Assessment Methodologies: A State of the Art Master's Thesis*. Universitat de Lleida; May 2011. Obtained from: http://repositori.udl.cat/bitstream/handle/10459.1/45831/Ali.pdf?sequence=2.
24. Life Cycle Assessment (LCA). *A Guide to Approaches, Experiences and Information Sources*. Denmark: EEA (European Environmental Agency); 1998. ISBN 92-9167-079-0.
25. Pennington DW, Potting J, Finnveden G, Lindeijer E, Jolliet O, et al. Life cycle assessment. Part 2: current impact assessment practice. *Environment International*. 2004;30:721–739.
26. Lindeijer E. *'Normalisation and Valuation', Part VI of the SETAC Working Group Report on LCA Impact Assessment*. The Netherlands: IVAM Environmental Research, University of Amsterdam.
27. Finnveden G. Valuation methods within LCA—where are the values? *International Journal of Life Cycle Assessment*. 1997;2(3):163–169.
28. Curran MA, Mann M, Norris G. The international workshop on electricity data for life cycle inventories. *Journal of Cleaner Production*. 2005;13(8):853–862.
29. Curran MA, Mann M, Norris G. *'Report Summary on the International Workshop on Electricity Data for Life Cycle Inventories', Held at the Breidenbach Research Center, Cincinnati, Ohio, 23–25 October 2001*. EPA/600/R-02/041; July 2002. Available at: http://nepis.epa.gov/Adobe/PDF/P1001NRO.pdf.
30. *Attributional vs. Consequential Life Cycle Analysis of Biodiesel*. Biodiesel TechNotes. Department of Biological and Agricultural Engineering, University of Idaho. Available at: http://web.cals.uidaho.edu/biodiesel/files/2012/11/ALCACLCA.pdf.
31. Brander M, Tipper R, Hutchison C, Davis G. *Consequential and Attributional Approaches to LCA: A Guide to Policy Makers with Specific Reference to Greenhouse Gas LCA of Biofuels*. Technical Paper. Ecometrica; April 2008. Available at: http://ecometrica.com/white-papers/consequential-and-attributional-approaches-to-lca-a-guide-to-policy-makers-with-specific-reference-to-greenhouse-gas-lca-of-biofuels/.

32. Thomassen M, Dalgaard R, Heijungs R, Boer I. Attributional and consequential LCA of milk production. *International Journal of Life Cycle Assessment*. 2008;13(4):339–349.
33. Goedkoop M, Hofstetter P, Müller-Wenk R, Spriensma R. The Eco-Indicator'98 explained. *International Journal of Life Cycle Assessment*. 1998;6(3):352–360.
34. Goedkoop M, Spriensma R. *The Eco-indicator'99, a Damage-Oriented Method for Life Cycle Impact Assessment*. Den Haag: VROM; 1999.
35. Hofstetter P. *Perspectives in Life Cycle Impact Assessment: A Structured Approach to Combine Models of the Technosphere, Ecosphere, and Valuesphere*. Boston: Kluwer Academic Publishers; 1998.
36. Goedkoop M, Oele M, de Schryver A, Vieira M, Sander H. *SimaPro Database Manual*. The Netherlands: Methods Library, PRe Consultants; 2010.
37. Goedkoop M, Spriensma R. *The Eco-indicator'99, a Damage-Oriented Method for Life Cycle Impact Assessment*. 3rd ed. The Netherlands: PRe Consultants, BV; 2001.
38. Earthshift. *Simapro Impact Assessment Methods, Eco-indicator'99*. Available at: http://www.earthshift.com/software/simapro/eco99.
39. Earthshift. *Simapro Impact Assessment Methods, CML 2001*. Available at: http://www.earthshift.com/software/simapro/clm2001.
40. Recipe Website. Available at: http://www.lcia-recipe.net/.
41. PRe Consultants. *Recipe Method*. Available at: http://www.pre-sustainability.com/recipe.
42. Goedkoop MJ, Heijungs R, Huijbregts M, De Schryver A, Struijs J, Van Zelm R. *ReCiPe 2008, 'A Life Cycle Impact Assessment Method Which Comprises Harmonised Category Indicators at the Midpoint and the Endpoint Level*. Report I: Characterisation. 1st ed. January 06, 2009.
43. Acero AP, Rodríguez C. *Andreas Ciroth, LCIA Methods- Impact Assessment Methods in Life Cycle Assessment and Their Impact Categories*. Green Delta; January 2015. Available at: https://www.openlca.org/wp-content/uploads/2015/11/LCA-METHODS-v.1.5.2.pdf. Accessed August 10, 2019.
44. European Commission. *List of Tools for LCA*. Available at: http://lca.jrc.ec.europa.eu/lcainfohub/toolList.vm.
45. Schepelmann P. *Life Cycle Assessment (LCA)*. Available at: http://www.ivm.vu.nl/en/Images/AT9_tcm53-161581.pdf.
46. Green Research, PE International. *Life Cycle Assessment – an Executive Overview of Applications, Market Drivers and Business Benefits*; July 2011. Available at: http://www.pe-international.com/fileadmin/Resources/Whitepapers_english/Whitepaper_-_LCA_Executive_overview_Green_Research.pdf.
47. Guinée JB, Gorée M, Heijungs R, Huppes G, Kleijn R, et al. *Handbook on Life Cycle Assessment – Operational Guide to the ISO Standards*. Dordrecht: Kluwer Academic, Dordrecht Publishers; 2002.
48. Ekvall T, Assefa G, Bjorklund A, Eriksson O, Finnveden G. What life-cycle assessment does and does not do in assessments of waste management. *Waste Management*. 2007;27:989–996.
49. Skone TJ. What is life cycle interpretation? *Environmental Progress*. 2000;19(2):92–100.

Life cycle assessment and product carbon footprint modelling of textile products

7

7.1 Introduction

Assessment of the environmental impacts for any product or process requires modelling of the system under study. This is a complex process involving many methodological issues. The complexity of modelling varies according to the type of life cycle assessment (LCA), e.g. cradle to gate, cradle to cradle or cradle to grave. Modelling of a product or process system for an LCA or product carbon footprint (PCF) study is a pivotal part of the procedure, with a significant impact on the results of the study. In this chapter, the modelling aspects of LCA and PCF assessment of textile and clothing products are discussed. Issues such as the methodological complexities of modelling, available databases, filling of data gaps and problems specific to modelling textile products will be highlighted.

7.2 Modelling for product carbon footprint and life cycle assessment of textile products

Previous chapters have discussed how environmental impacts can be measured for products/processes by different assessment techniques and methods. LCA is one such technique, which is both widely known and commonly used. Any indicator, for example, carbon footprint, water footprint or ecological footprint, or multiple indicators of a product, process or service system can be quantified using the LCA approach.

Once a decision has been made to perform a product-based or process-based LCA, further steps towards modelling can begin. Product-based LCAs will be addressed here first. The initial step in modelling is to confirm the stages of life cycle to be included, the boundaries of the study, and so on, as discussed in Chapter 6 on the first phase of LCA, i.e. goal and scope definition. At this point, many details have to be confirmed, such as the functional unit and reference flows of the study, along with its scope and boundaries. A detailed LCA study should adhere to the ISO 1404X series of standards for LCA and the four steps detailed therein must be strictly followed (as discussed in Chapter 6).

A product-based LCA study may only include cradle to gate stages, or can be extrapolated to cradle to grave or even cradle to cradle stage. It should be noted that there are many other variants of an LCA study and this chapter mentions only a few of the most important. Based on the selected variant, life cycle phases of a textile product will vary, as listed in the following:

Assessing the Environmental Impact of Textiles and the Clothing Supply Chain
https://doi.org/10.1016/B978-0-12-819783-7.00007-7

- raw materials manufacturing stage, including growth, acquisition and processing;
- fibre production processes;
- yarn, fabric and garment production processes;
- packaging of the garment with accessories and labels;
- distribution;
- retailing;
- consumer use phase;
- end-of-life management, including reuse, recycling, remanufacture, incineration with or without energy recovery and disposal at landfill.

Life cycle stages of different variants are illustrated in Fig. 7.1.

Having defined goal and scope, it is useful to draw the process flow diagram, including all details pertaining to the product under study. Figure 7.2 gives a generalized picture; this is quite subjective and will vary from case to case.

Depending on the stages of life cycle to be included, process flow and data collection details will vary. Once the details of goal and scope definition, such as boundary, reference flow and functional units, have been finalized, the next step is to gather the inventory details for the product system under study. This step is often regarded as time consuming and tedious. Each life cycle phase in the product system comprises many elements and sub-elements; the inventory details of each of these elements must be modelled accordingly. All elements in a product system should be covered according to the details defined in phase 1.

The ISO 1404X series of LCA standards dictate that life cycle inventory (LCI) data must be collected for each unit process, i.e. the smallest portion of a product system for which data are collected.[2] Modelling and calculation of the inventory results are the core activities of the LCI analysis step.[3] Compiling an inventory of all the processes involved in the life cycle of the product system(s) under study, detailing all relevant interactions with the environment, is the chief focus of the inventory analysis phase.[4] Each life cycle phase in the product system should be broken down into individual unit processes; the data for each unit process are then collected, as shown in Fig. 7.3. Modelling of the product system identifies the unit processes for which data are required.

An LCA software tool can be employed to model the product system and to quantify the LCI results. For simplified LCA studies, it may be sufficient to use an

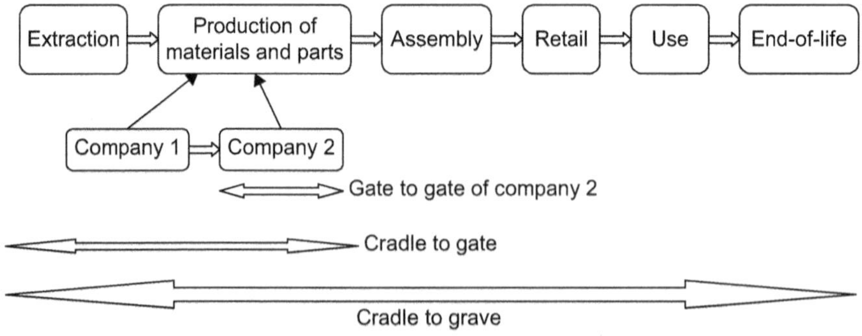

Figure 7.1 Life cycle stages of different variants.[1]

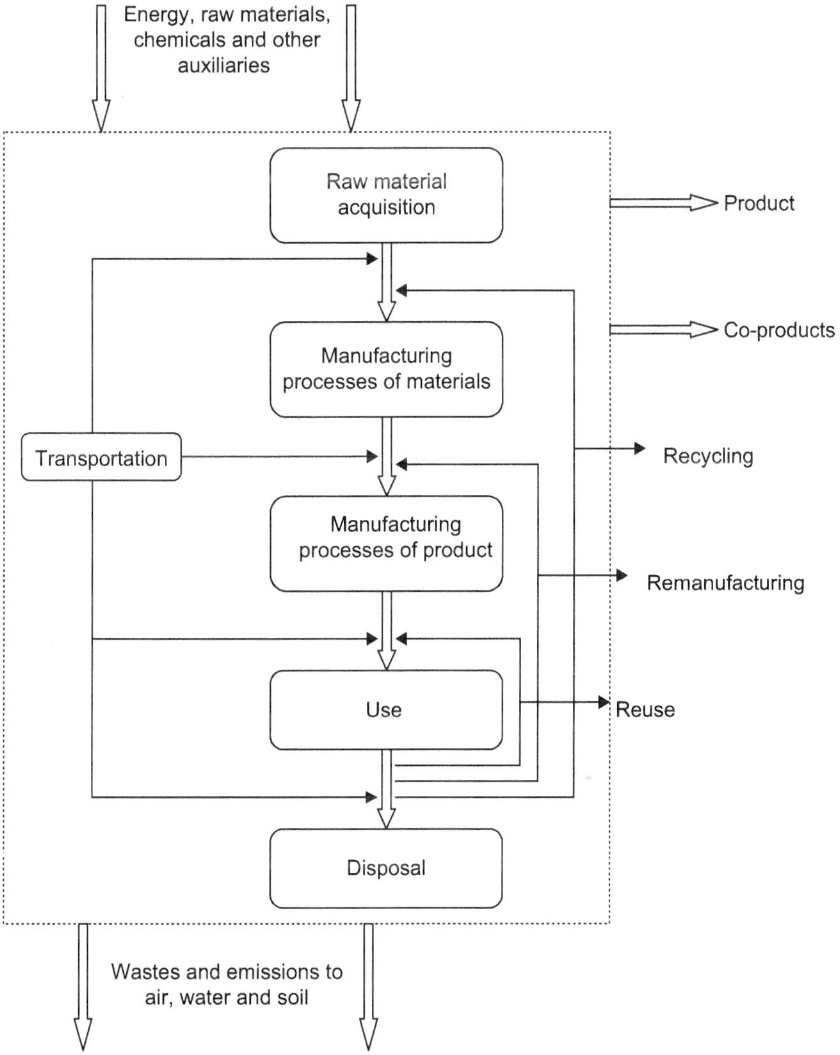

Figure 7.2 Process flow diagram.

electronic spreadsheet program, such as Microsoft Excel.[3] Detailed steps associated with LCI are listed in the following (most of these were discussed in Chapter 6).

- Data collection – for all unit processes and for all life cycle phases associated with the product system.
- Data validation – establishment of the data.
- Relating data to unit process – quantitative input and output of the unit process is calculated according to the reference flow.
- Allocation – inventories are dealt with according to products and co-products (as explained previously).

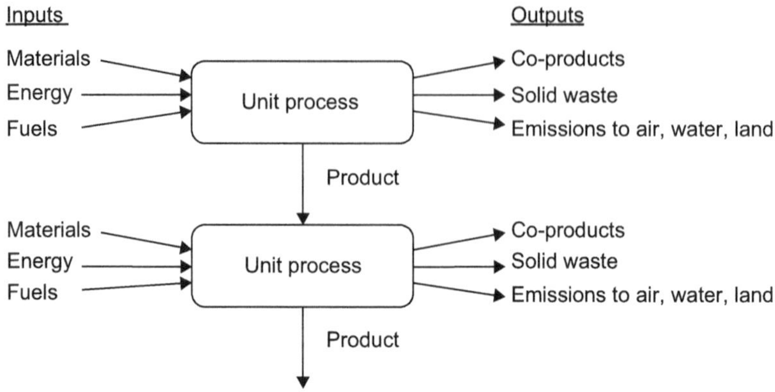

Figure 7.3 Unit processes and data collection for life cycle inventory.[3]

- Data aggregation – normalization of the flows of all unit processes in the system to the functional unit.
- Refinement of system boundaries, as required.[2,3]

Data collection is the most difficult and significant phase of LCI because it directly influences the final result. Hence, a high level of expertise is needed in the LCI modelling. For any product system, there are two types of data: primary and secondary.

Primary data are obtained through direct measurement at the site or through calculation based on direct measurements at the original source (site-specific data). Primary data represent the best choice and should be collected on-site by the LCA practitioner for the various elements that constitute the complete life cycle phases of the product in question. For this to be achieved, a form of systematic data collection for the product system must be created by the practitioner, which can be used by the suppliers to complete the required data.

Secondary data are collected through secondary data sources, i.e. by means other than direct measurement. Such data collection is encouraged when primary data collection is not possible for whatever reason; the reason/s should be justified clearly in the LCA report. An LCA practitioner can obtain data from various secondary data sources, such as previously published LCA reports, scientific publications, publicly available databases, government papers and reports, statistical reports, surveys and interviews with those concerned. Authenticity, time limit and geographical coverage of secondary data sources are all aspects that need to be carefully examined before using such data from an earmarked source.

The best approach is to gather as much primary data as possible and then fill in the data gaps with the aid of secondary data. For an LCA study, a practitioner may seek data from secondary sources for certain processes (depending on the availability of primary data), such as

- extraction and production of raw materials;
- production of ancillary materials, such as chemicals or packaging materials;
- production of fuels;

- production of energy from different sources;
- transportation;
- wastewater and waste treatment.

Unit processes should be compiled for each life cycle phase; finally, all the life cycle phases must be aggregated for the product system under study. This completes the model of the product system. Once a model has been created for the complete life cycle phases of a product under the defined boundaries and scope of the project, any impact category can be studied from that model, for example, carbon footprint, ecological footprint, eco-damage indicators or other indicators such as human toxicity, eutrophication, or acidification, from various impact assessment methods.

7.3 Available databases for life cycle assessment and product carbon footprint modelling of textiles and the clothing supply chain

As previously mentioned, the biggest challenge or handicap faced by textile researchers working in the field of environmental and sustainability assessments is the availability of a fully fledged, dedicated database for textiles and the clothing supply chain. Data collection from primary sources is the best method, and then certain data gaps can be filled with the aid of secondary data sources, as discussed at the end of Section 7.2. This holds true for full LCA studies, which are labour and resource intensive, are time consuming and demand a high level of expertise and precision.

At times, it may be necessary to look into the life cycle impacts of various textile materials at different points of the supply chain. In such cases, a screening or quick LCA should be used, depending entirely on data from databases of any LCA software used or other secondary data sources, as discussed earlier. The greatest difficulty in this situation is data availability for textile materials/textile processes (or other elements of the textile supply chain) or for the modelling of certain life cycle phases, such as use and disposal scenarios.

LCI databases for textiles do exist, with some that are publicly available, such as Ecoinvent, GaBi and ESU LCI database. LCI data relating to T-shirts made from cotton and synthetics, and their retail distribution, are available in the textile materials category of the ESU LCI database; these data have to be purchased with associated cost.[5]

Ecoinvent and GaBi databases have several datasets pertaining to textiles and the clothing supply chain. The GaBi database has extensive datasets for textile finishing, consisting of 147 processes and 140 plans for different areas of textile finishing. These include the following:

- Pre-treatment
  - dry processes such as singeing
  - wet processes such as desizing, bleaching and scouring.
- Dyeing and/or printing (e.g. acids, cationic, direct, disperse and reactive dyes).
- Finishing, fabrics.[6]

Detailed explanations of the different datasets available in textile finishing can be found from the GaBi webpage.[7] GaBi also has other datasets pertaining to textiles, including cotton fibres from recycling of clothes, flax and hemp fibres and sisal fibres under the renewable raw materials category.[8]

Similarly, the Ecoinvent database has various datasets pertaining to textile materials and processes. In particular, the most recent version of Ecoinvent database (V3) has many additions in these areas. Various datasets available in Ecoinvent pertaining to textiles from fibre formation to fabric formation, including some examples (these are available from the search option of the Ecoinvent website using different key words[9]; most of the textile materials and processes are available in Ecoinvent Report No. 21[10]).

- Cotton production – 1 kg of cotton fibre as a reference product for China, the United States, and the global (GLO) average.
- Cotton seed production, for sowing – 1 kg of cotton seed, for sowing, as a reference product for the United States and the GLO average.
- Jute production, irrigated – 1 kg of jute fibre as a reference product for India and the GLO average.
- Jute production, rainfed – 1 kg of jute fibre as a reference product for India and the GLO average.
- Kenaf production – 1 kg of kenaf fibre as a reference product for India and the GLO average.
- Polyester resin production, unsaturated – 1 kg of polyester resin, unsaturated as a reference product based on the European average data and the GLO average.
- Nylon 6.6 production, unsaturated – 1 kg of Nylon 6.6 as a reference product based on the European average data and the GLO average.
- Nylon 6 production, unsaturated – 1 kg of Nylon 6 as a reference product based on the European average data and the GLO average.
- Polypropylene production, granulate – 1 kg of polypropylene production, granulate as a reference product based on the European average data and the GLO average.
- Textile production, jute – 1 kg of jute textile as a reference product for India and the GLO average.
- Textile production, kenaf – 1 kg of kenaf textile as a reference product for India and the GLO average.
- Viscose fibre production – 1 kg of viscose fibres as a reference product for the GLO average.
- Yarn production for cotton fibres for the GLO average.
- Yarn production for kenaf and jute fibres for India.
- Textile refinement for the GLO average, which includes bleaching, washing, drying and dyeing processes for cotton lint, yarn and fabric.
- Textile weaving process for bast fibres for India, and the weaving process for cotton fibres for the GLO average.
- Textile spinning process for bast fibres for India, and spinning process for cotton fibres for the GLO average.
- Textile production, knit cotton, batch dyed – 1 kg of textile, knit cotton as a reference product for the GLO average.
- Textile production, knit cotton, yarn dyed – 1 kg of textile, knit cotton as a reference product for the GLO average.

- Textile production, woven cotton – 1 kg of textile, woven cotton as a reference product for the GLO average.
- Treatment of waste textile, soiled, municipal incineration – 1 kg of solid textile waste as a reference product for China and the GLO average.

A few new datasets have recently become available in Ecoinvent.

- ✓ Cottonseed meal to generic market for protein feed – GLO average
- ✓ Cottonseed oil, refined, to generic market for vegetable oil, refined – GLO average
- ✓ Market for cotton fibre – GLO average
- ✓ Market for cotton seed – GLO average
- ✓ Market for irrigation – India
- ✓ Market for textile, knit cotton – GLO average
- ✓ Market for waste textile, soiled – GLO average
- ✓ Spinning bast fibre – India

Apart from Ecoinvent and GaBi, there are other databases with datasets pertaining to textiles and the clothing sector. For instance, IDEMAT (developed by the Delft University of Technology) has datasets pertaining to textiles for the following elements:

- production of cotton fibres (in the United States);
- production of cotton fabric (bleached, dyed cotton apparel in the United States);
- production of polyester fabric (dyed, manufactured polyester apparel, based on average data);
- production of eco-cotton fabric (eco-cotton in the United States).

IDEMAT datasets are available in commercial LCA software tools, such as SimaPro, but have not been updated recently.

7.4 Key issues in using databases

Databases are an important element in an LCA study because the success of any such study largely depends on the quality of the data used (i.e. reliability and accuracy). Another important issue is the availability of suitable data. These two main issues are currently pertinent to LCA databases for textiles and the clothing sector, apart from other methodological complexities that will be dealt with in the next section.

As discussed earlier, either site-specific (primary) or secondary data can be used in an LCA study. Even when primary data are available, secondary data from various reliable sources will be needed to fill any data gaps. For textiles and the clothing sector, availability of secondary data sources and their suitability is a major challenge for LCA practitioners. Publicly available databases such as Ecoinvent and GaBi have some datasets pertaining to textile materials and processing, but these are not complete. The entire supply chain cannot be mapped with the datasets currently available. The second issue is data suitability. The listed textile datasets mostly pertain to Europe or the United States, or represent a GLO average. However, most textile production is based in Asia, where publicly available datasets are very scarce.

LCA studies conducted in India or Bangladesh, for example, therefore have to be built using datasets that represent European conditions or US practices. Clearly, this is awkward in view of the very wide differences between different regions of the world in terms of aspects such as working conditions, state of the art of machines used for production, pollution abatement measures, governmental laws governing pollution, availability and production of energy, transportation practices, differences in energy sources, sources of water and the fate of different pollutants released into the atmosphere. This lack of relevance is also clearly apparent to any would-be users of the study. The need to rely on data that are not strictly applicable in terms of geographical location, technological limitations and methodological limitations is therefore a major challenge to the LCA practitioner.

Sometimes the data for all the life cycle phases defined for a textile product cannot be obtained from the same database source. Although not necessarily a generalized or widespread issue, this situation is common for textile products. Obtaining data from different data sources for a study presents several challenges for the practitioner to overcome. A complicating factor is that not all databases are updated in a timely manner. Using a database that has not been updated for the past 10 years, as an example worst-case scenario, makes it very difficult for the LCA practitioner to justify any results obtained.

7.5 Difficulties in modelling and simulation

Difficulties in modelling and simulation of various life cycle impacts for different indicators have been discussed in previous chapters. Some major difficulties encountered by LCA practitioners in modelling textile materials and products for LCA are listed in the following:

- Textile products have a very long supply/production chain: Many elements need to be considered and modelled in terms of any LCA study.
- Definition of a suitable functional unit for cradle to gate and cradle to grave stages is difficult. Modelling of cradle to grave stages, involving both consumer use and disposal phases, is highly complex. For textile products, the functional unit must be defined incorporating the use stage in cradle to grave assessments. For example, it is not helpful to study the life cycle impacts of 1 kg of a T-shirt or 1 kg of cotton trousers. Instead, the life cycle study should concentrate, for example, on quantifying the impacts of 3 years of T-shirt use in one individual, or 5 years of cotton trouser use. Also, care should be taken when making comparative studies of different textile materials because each fibre has its own washing and durability characteristics.
- Operational boundaries in the goal and scope definition: These must be dealt with carefully. Inclusion or omission of areas of consideration should be justified properly in the report.
- Data choices, assumptions and limitations are unique to each study: These should be carefully managed and full information revealed to the reader/user of the LCA study.
- Availability of datasets and filling of data gaps: It is always very difficult to source the required data of sufficient quality at all times. As previously mentioned, there is a dearth of publications/reports/databases dedicated to the LCA of textile and clothing products.

- Conversion/acquisition of data from other studies/sources for any specific study: Every study has its unique data sources, boundaries, assumptions and limitations; therefore, converting data for application in a different situation demands a high level of expertise in the subject.
- Dealing with data uncertainty: This is always challenging in LCA studies and textile products are no exception, owing to their lengthy supply chain.
- Obtaining detailed information: When referring data from other sources/studies, it is often found that they deal with the totality of any element and do not spell out any detail regarding the various sub-processes. For example, water and energy use tend to be reported in total for a product's life cycle, and not in terms of any individual processes in that life cycle. This presents a difficulty when trying to understand and acquire data for a particular process. The same holds true for the type of water/energy source from which water was obtained, or the source of wastewater discharge. These would all be usually reported in detail in LCA studies.
- Confidentiality: Most LCA studies do not disclose detailed information with respect to LCI. The practitioner therefore has to deal with a transparency issue in LCA studies.[11]
- Suitability of data obtained from secondary sources: For example, when a study refers to the LCA of denim trousers produced in Bangladesh, but the details of packaging are obtained from the Ecoinvent database (which represents conditions in Europe), the data are clearly inappropriate. However, it is often necessary to deal with such limitations in an LCA study.
- Definition of impact categories suitable to the product under study can be tedious: This demands a high level of expertise in the subject and the choices of impacts must be justified in the report.
- Allocation: This is one of the most difficult parts of LCA modelling of any product, including textiles. One textile example is the allocation of resources and emissions between meat products and wool fibre for wool textiles; similarly, allocation between cotton seed and cotton fibres in the case of cotton textiles. Issues surrounding economic/physical allocation and subsequent modelling represent a major challenge in textile-based LCA studies. The choice of allocation method chosen for an LCA study should be fully justified. Also, for certain products or elements, the allocation procedure can be a tedious job.
- Modelling of electricity-based energy inputs: This demands various details such as the geographical location of the product's manufacturing phases, sources of electricity from that location, ways of manufacturing electricity at that location, the transportation factor and the transmission losses. In most cases, the electricity mixes prevailing in that location must be used for modelling, and obtaining the electricity mix details from a reliable source is tedious. Similarly, it is also difficult to model obtaining energy via other, different energy carriers. Finally, obtaining datasets for energy elements in Asian countries is not feasible using the current LCI databases.
- Inclusion of the effects of dust and noise: These are an inevitable consideration in relating textile manufacturing with the current LCA practice and methods.
- Modelling use and disposal phases of textiles and clothing products: This involves consideration of consumer behaviour, which varies greatly between individuals. Incorporating geographical location in the model is also important (i.e. where the product is going to be used) because use and disposal phases vary between different parts of the world. The greatest difficulty lies in obtaining a real scenario (as relying on assumptions can be misleading). Consumer surveys can help develop an understanding of the use and disposal behaviour of the product in question, in order to model accordingly. Administering a questionnaire survey via any media, however, is very demanding in terms of time and resources.

- Quantification of reuse and recycling (closed-loop systems in particular) of textile products and allocating the benefits via life cycle models is a complicated process, which demands a high level of modelling expertise and subject knowledge.

7.6 Summary

One aspect of quantifying the environmental impact of a textile product is to conduct an LCA or PCF study by preparing the life cycle model of that particular product. Life cycle models need to cover all the life cycle phases of the product under consideration. There are two important variants of LCA models that are widely used in this respect, namely, cradle to gate and cradle to grave.

Modelling of the life cycle phases using the guidelines provided by ISO 1404X standards for textile products poses various challenges and difficulties to textile LCA practitioners. This chapter has discussed ways to create models for textile products for different variants and how to fit these models with appropriate data for the various life cycle phases that need to be considered.

Although only a few databases and datasets are available for different textile and clothing products, popular databases such as Ecoinvent provide some good datasets. This chapter has highlighted suitable sources of such data for modelling textile and clothing products and has discussed some of their limitations. The various challenges and difficulties of modelling LCA or PCF study of textile and clothing products have also been described.

7.7 Sources of further information and advice

- Baumann H. *Life Cycle Assessment and Decision Making – Theories and Practices* [Ph.D. thesis]. Technical Environmental Planning, AFR Report 183. Göteborg, Sweden: Chalmers University of Technology; 1998.
- Beck A, Scheringer M, Hungerbühler K. Fate modeling within LCA – the case of textile chemicals. *International Journal of LCA*. 2000;5(6):335–344.
- Consoli F, Allen D, Boustead I, Fava J, Franklin W, et al. eds. *Guidelines for Lifecycle Assessment: A 'Code of Practice'*. Brussels: Society of Environmental Toxicology and Chemistry (SETAC); 1993. 69 pp.
- Heiskanen E. The institutional logic of life cycle thinking. *Journal of Cleaner Production*. 2002;10:427–437.
- Dahllöf L. *Life Cycle Assessment (LCA) Applied in the Textile Sector: the Usefulness, Limitations and Methodological Issues – A Literature Review*. Environmental Systems Analysis, ESA- R2003:9, ISSN 1404-8167. Göteborg, Sweden: Chalmers University of Technology; 2003.
- Neitzel H. Principles of product-related life cycle assessment – 'Conceptual framework'/ 'Memorandum of understanding'. State October 1993. *International Journal of Life Cycle Assessment*. 1996;1(1): 49–54.
- *Nordic Guidelines on Life-Cycle Assessment*. Nordic Council of Ministers. Nord; 1995. p. 20.
- Frischknecht R, Rebitzer G. The ecoinvent database system: a comprehensive web-based LCA database. *Journal of Cleaner Production*. 2005;13:1337–1343.

- SETAC 1993. *Guidelines for Life-Cycle Assessment: A 'Code of Practice'*. 1st ed. From the SETAC Workshop held at Sesimba, Portugal 31 March – 3 April. Society of Environmental Toxicology and Chemistry.
- Tukker A, Huppes G, Guinée J, Heijungs R, Koning A, Oers L, Suh S. Environmental Impact of Products (EIPRO) analysis of the life cycle environmental impacts related to the final consumption of the EU-25. EUR 22248 EN; 2006.

7.8 References

1. European Commission – Joint Research Centre – Institute for Environment and Sustainability. *International Reference Life Cycle Data System (ILCD) Handbook – General Guide for Life Cycle Assessment – Detailed Guidance*. EUR 24708 EN. 1st ed. Luxembourg: Publications Office of the European Union; March 2010.
2. International Standards Organization. *Environmental Management – Life Cycle Assessment – Principles and Framework. ISO 14040*, 2006-07-01. 2nd ed. 2006.
3. Koskela S, Hiltunen M-R. *A Guide to the Utilization of LCI/LCA Databases for Estonian Enterprises*. Helsinki: Finnish Environment Institute, Mimeograph Series of Finnish Environment Institute; 2004. Available at: http://www.ymparisto.fi/download.asp?contentid=26849.
4. Guinée JB, Gorée M, Heijungs R, Huppes G, Kleijn R, et al. *Handbook on Life Cycle Assessment – Operational Guide to the ISO Standards*. Dordrecht: Kluwer Academic, Dordrecht Publishers; 2002.
5. ESU Services. *ESU Life Cycle Inventory Database on Demand*. Available at: http://esu-services.ch/data/data-on-demand/ [accessed on 15 August 2019].
6. GaBi Software. *GaBi Databases, Textile Finishing*. Available at: http://www.gabi-software.com/databases/gabi-databases/textile-finishing/ [accessed on 16 August 2019].
7. GaBi Software. *GaBi Databases, Textile Finishing*. Available at: http://www.gabi-software.com/support/gabi/gabi-database-2019-lci-documentation/extension-database-xv-textile-finishing/ [accessed on 16 August 2019].
8. GaBi Software. *GaBi Databases, Renewable Raw Materials*. Available at: http://www.gabi-software.com/support/gabi/gabi-6-lci-documentation/extension-database-xii-renewable-raw-materials/ [accessed on 17 August 2019].
9. Ecoinvent Centre. *Swiss Centre for Life Cycle Inventories, Database Search*. Available at: https://ecoquery.ecoinvent.org/Search/Index.
10. Althaus H-J, Werner F, Settler C, Dinkel F. *Life Cycle Inventories of Renewable Materials*. Ecoinvent Report, No. 21. Swiss Centre for Life Cycle Inventories; 2008.
11. Dahllöf L. Methodological Issues in the LCA Procedure for the Textile Sector – A Case Study Concerning Fabric for a Sofa. *ESA – Technical Report 7, Environmental Systems Analysis*. Sweden: Chalmers University of Technology; 2004.

End-of-life management of textile products

8

8.1 Introduction

The range and volume of textile products used every day is growing exponentially and so is the problem of their disposal. Traditionally, disposal meant sending unwanted products to landfill. However, the volume of waste, the diminishing amount of new land available for landfill, concern about pollution and health issues caused by landfill waste, and a growing awareness of the need to conserve scarce resources have meant searching for more sustainable and environmentally friendly alternatives.

The textile sector is characterized by many disposable products (e.g. wipes or baby diapers) with a short life cycle. The rapid development of the fashion industry and the move to shorter, 'fast' fashion cycles have made the problem worse by shortening the life of many fashion items, which may now be discarded by consumers very rapidly in response to changing fashion trends even if the product is, in all other respects, still wearable. Any product, including textile products, creates multiple environmental issues over its entire life cycle from cradle to grave. However, the end-of-life phase requires particular attention.

Because of the increasing rates of population growth and urbanization, consumption of textile products is increasing day by day, exacerbating the problem of disposal. According to statistics from the US EPA, around 5% of space in landfills is occupied by textile waste. Although in recent days recycling is getting popular, still the amount of waste getting recycled is not up to the mark. As per the estimation of US EPA, the textile recycling industry recycles approximately 3.8 billion pounds of post-consumer textile waste (PCTW) each year, this only accounts for approximately 15% of all PCTW, leaving 85% in our landfills. Dealing with this huge amount of piled up waste in landfills and reducing it has become a hot topic in the field of sustainability of textiles and clothing. There are many new innovations and developments in this sector, most of them focusing on zero waste. Eco-design of textiles is another innovative area, which promotes the creation of products that last longer, can be reused/recycled and are biodegradable. Such efforts combine towards achieving sustainability in the textiles and clothing sector, with a principal focus on diverting such consumables from landfill disposal.

A product can be termed 'green' if it creates low environmental impacts throughout its life cycle and generates the lowest possible or nil environmental impacts at its end of life. The crux of the eco-design concept is that these qualities are demonstrated throughout a product's journey from inception to end of life; this should be borne in mind by researchers involved in developing green/sustainable textile products.

This chapter is dedicated to the end-of-life management of textiles and clothing products. Various approaches to the problem are discussed, alongside the implications associated with reuse, recycling, incineration, landfilling and biodegradation.

Assessing the Environmental Impact of Textiles and the Clothing Supply Chain
https://doi.org/10.1016/B978-0-12-819783-7.00008-9

8.2 End-of-life product management options

When it leaves the manufacturer's gate, a textile product enters the retail and distribution channels, via which it then reaches the customer. From this point, the use phase begins. After a certain period, the customer decides to dispose of the product and this is the end-of-life phase. There are many destinations or fates of a product at the end of its life. These include

- reuse for primary and secondary purposes,
- recycling (open- and closed-loop types),
- landfilling,
- incineration.

From the perspective of life cycle assessment (LCA) or environmental impact assessment, two terms can be applied at this point: one is 'impact' and the other is 'benefit' or 'credit'. These terms are self-explanatory; clearly the end-of-life option that brings credits or benefits is favoured over one that causes any impacts. Each of the options for disposal has its own pros and cons, and these will be discussed in the sections that follow, with a focus on textile products.

Reuse is the first and best option in terms of gaining only environmental benefits or credits with nil impacts. Recycling is the next best option, and it brings both benefits and certain impacts, followed in the hierarchy by incineration. Incineration can be performed with or without energy recovery. Clearly, incineration with energy recovery is preferred (generating impacts as well as bringing benefits). The final and most undesirable option is disposal in a landfill, which creates environmental impacts only. The perceived degree of these impacts and benefits is quite subjective and cannot be generalized, as it is influenced by multiple factors. This aspect is not covered in the current discussion.

To summarize, as defined by the 2008 European Union Waste Framework Directive (European Council, 2008), waste prevention is the best choice of waste management, followed by the other options in the hierarchy of reuse—recycle—energy recovery—landfill.[1]

Owing to the limitations of space in landfills and also to promote the concepts of resource conservation and to minimize the environmental impacts of producing a new product, waste management options and concepts are being refined. A very famous concept called 3R's — Reduce, Reuse and Recycle — has been refined to 5R's as Refuse, Reduce, Reuse, Repurpose and Recycle in the same order of hierarchy, as keeping the same order is also crucial. Similarly, the new hierarchy of waste management has been known as the ARRE strategy: Avoid, Reduce, Recycle and Eliminate.

8.3 Reuse of textile products

As the name implies, reuse of a product simply means using it again to delay its eventual disposal. This can either take the form of primary reuse, which is reuse for the same purpose, or secondary reuse, which is reuse for some purpose other than that originally intended. Reuse can be achieved in many ways, but principally there are three types:

1. formal reuse (e.g. secondhand stores)
2. semi-formal reuse (e.g. selling through websites such as Ebay, Blocket and so on)
3. informal reuse (e.g. when children inherit clothes from siblings, or sharing amongst friends and relatives).[2]

Formal reuse is principally carried out by charity organizations, who collect secondhand clothes and allow them to be reused by people in need. In most countries, it is common to see used clothes bins where people can place their unwanted clothes, knowing that they will reach someone who needs them.

People can sell/buy secondhand clothes for reuse through some websites. Such semi-formal reuse is a huge and profitable business for websites such as Ebay and Blocket.[2]

Informal reuse is commonplace in any home, where younger children get the hand-me-down clothes from their older siblings for reuse. Baby clothes are reused in the same way, either in the same household or via relatives. This sort of reuse is most common for baby clothes because they are so quickly outgrown. Similarly, adult clothes are reused by sharing amongst friends and relatives.

Reuse brings ample benefits to the environment and remains a highly preferred option, as it does not demand any resources or emit any pollutants in order to provide that benefit, so there is no associated impact. This statement holds true for direct reuse scenarios, but there may be opportunities for reuse of textile products via certain preprocesses such as collecting, processing and transporting items. In this scenario, there may be environmental impacts, but these are generally outweighed by the benefits accrued from reuse.

The reuse of clothes provides benefits in two main ways:

1. it delays disposal of the current product, preventing it from reaching landfill too soon;
2. it minimizes the production of new textile products with their associated environmental impacts.

Previous studies have noted that the benefits to the environment from reduced production (avoided burdens) are far greater than for recycling or energy recovery.[2–4] Apart from the other environmental benefits, the reuse option lowers the carbon footprint (global warming potential [GWP]) by 15 kg CO_2 equivalents per kilogram of textiles.[2,4,5] As discussed in detail in previous chapters, virgin textile production is energy intensive and emits various pollutants to air, water and land and is consequently responsible for creating various environmental impacts. Aside from this, it is challenging to balance the equation of shrinking resources with a growing population. Reuse helps address such issues to a great extent.

Such findings have been confirmed by quantitative studies as well.[6–11] According to a study conducted by Farrant in 2008,[11] reuse processes such as collection, processing and transportation of secondhand clothing create insignificant environmental impacts when compared with the actual benefits gained through replacing the equivalent quantity of virgin products. This study quantified the impacts/benefits of 100 T-shirts and 100 65% polyester/35% cotton trousers entering the secondhand clothing business as compared to their direct disposal. In the case of T-shirts, the following benefits were estimated:

- 14% reduction in global warming,
- 28% reduction in acidification impacts,
- 25% reduction in nutrient enrichment impacts,
- 15% reduction in natural gas and crude oil consumption,
- 24% saving for human toxicity in soil,
- 30% reduction in waste generated.[10]

Similarly, for the 100 pairs of trousers made from 65% polyester/35% cotton, the following benefits were estimated:

- 23% reduction in global warming,
- 28% reduction in acidification impacts,
- 25% reduction in nutrient enrichment impacts,
- 20% reduction in natural gas and crude oil consumption,
- 30% saving for human toxicity in soil,
- 45% saving for human toxicity in water,
- 25% reduction in waste generated.[10]

Owing to the popularity of fast fashion in the recent days, consumers have the choice and tendency to update their clothing products in their shelves. This leads to an important issue that adds to the waste management problem of textile products, which is shortening of the practical service lives. There are many attempts across the industry to alleviate the environmental impacts of fast fashion and one of them is collaborative consumption. Clothing library is one of the ways to practice collaborative consumption. A LCA study was done on the subject to compare the environmental impact of clothing libraries and conventional fast fashion business models. Three products were part of this study's assessment, namely, a T-shirt, jeans and dresses, and from the study, it was concluded that there are benefits rather than impacts if clothing libraries were implemented, which is associated with the prolonged service lives of garments. Another important finding to be noted from this study is the benefits gained from reduced production in terms of collaborative consumption can be completely offset by the consumer transportation. Hence, one has to carefully account for the logistics part while implementing the collaborative consumption business models.[12]

8.4 Recycling of textile products

The next best option in terms of waste management is recycling. Recycling refers to the deconstruction of a discarded product after use and further processing of the resulting components to create either a similar or a different product (secondary product). This means that virgin materials, which would have been used to create the same product, are saved and also that landfill disposal is avoided/reduced, with the associated environmental benefits this method brings.

Previous chapters have described how recycling can be divided into two categories: open-loop and closed-loop recycling. Within a product's life cycle, if a material is recycled and rejoins the same life cycle without leaving it (e.g. as a raw material for the same product), then this is termed closed-loop recycling. However, if at the end-of-life stage, the material from a product is recycled and emerges as a new product altogether, leaving the current life cycle, then this is called open-loop recycling.

Recycling can also be divided into two categories according to the state of the material when it arrives for recycling: post-consumer and post-industrial waste recycling. As the name implies, if waste arrives for recycling after being discarded by the consumer at the end of its potential life, then this is called post-consumer waste recycling. On the other hand, if the material reaches the recycling point after it has been discarded at the industrial gate (without reaching the consumer phase at all), then it is called post-industrial waste recycling. Both types are common in the textile industry.

Recycling is an indispensable and highly desirable option for waste management. Certainly, reuse is the best option in the waste management hierarchy, preventing the occurrence of waste. After reuse, however, recycling is the most appropriate and desirable option. Even products selected for reuse must be recycled at some point in time, when their functional limit has been reached. On reaching that stage (end of actual life), they must be forwarded to recycling rather than disposed of at landfills in order to gain environmental benefits. Furthermore, not all products at the end of their life can be reused as secondhand clothes. Only a certain percentage may have the potential to be reused in this way, whereas others must be dealt with by the other waste management options; this is when recycling presents the best option.

Recycling is highly topical in a world where natural resources are becoming scarcer and environmental concerns are escalating over energy and water demands, as well as emission of pollutants through virgin product production. Recycling is also one way to balance shrinking resources against a growing population. Recycling of textile waste to replace virgin raw materials helps prevent depletion of non-renewable resources (raw materials for the production of synthetic fibres and even natural and renewable fibres demand non-renewable sources of energy for their production) and lowers the raw material cost. Furthermore, dependence on virgin raw materials, which are subject to various market influencing factors, is addressed by means of recycling.

The environmental implications of producing virgin textile materials have already been discussed in terms of

- energy needs;
- water needs;
- transportation requirements;
- requirement for various chemicals and auxiliaries;
- need for fertilizers, herbicides and pesticides;
- need for crop land;
- deployment of fossil fuels, which are non-renewable;
- a wide range of environmental impacts, including carbon footprint, acidification, nutrient enrichment, depletion of fossil fuels, ecological footprint, etc.

Previous chapters have discussed the amount of energy, water and other resources and the amount of various pollutants emitted, as well as other environmental impacts pertaining to different textile fibre production. Textile processes involved in producing the finished textile product have also been addressed. But to reiterate with an example here, production of 1 ton of baled cotton fibres requires around 46 GJ of energy (usually between 36 and 55 GJ/ton) and 5730 m^3 of water; production of 1 ton of cotton fibre emits 2 ton CO_2 equivalents.[13] The total amount of energy required to produce 1 ton of cotton T-shirts (approximately 4000 pieces) ranges from 188 to 325 GJ.[11,13,14] In this total apparel production expenditure of 325 GJ/ton of T-shirts, fibre production alone accounts for 55 GJ/t. These numbers show that potential exists for considerable savings when recycled fibres/apparel replace their virgin counterparts.

As mentioned earlier, the recycling operation demands a certain amount of resources, and this inevitably results in environmental impacts. Factors to be considered are the demand for energy during the mechanical or chemical processes involved, the requirement for additional raw materials and auxiliaries, as well as emissions into air, water and soil.[15,16] However, even considering these challenges, overall benefits can be derived from the process of recycling. The energy requirements and pollutant emissions from recycling processes are no more, and often much less than, those involved in production using virgin materials.[16]

Many studies dealing with the recycling of textile products have demonstrated this point. Results from case studies are worth mentioning here. In 2006, Allwood and coworkers[14] found that 1 ton of 100% cotton T-shirts (approximately 4000 pieces) and 1 ton of 100% viscose blouses (approximately 5000 pieces) demanded 188 and 235 GJ of energy for their production, whereas the collection and sorting of these quantities of T-shirts and blouses used only 6 GJ of energy in each case. Significant energy savings of 182 and 229 GJ/ton, respectively, were therefore achieved. Other studies have calculated the energy requirements for collection and grinding of textile materials, compared with the production of virgin materials.[14,17] Energy demand for collection and grinding of 1 ton of textile products was found to be 8GJ per ton, whereas the manufacturing effort to produce a ton of any fibre is significantly higher.

Another case study conducted to ascertain the carbon footprint of textile recycling in Sweden also confirmed the potential environmental benefits of recycling. In this LCA study, which was done in 2014, three different recycling techniques were considered for a model waste (as there were no recycling options present in Sweden when this study was done) consisting of 50% cotton and 50% polyester. The recycling processes included in this study were material reuse of textile waste of adequate quality, separation of cellulose from polyester using *N*-methylmorpholine *N*-oxide as a solvent and chemical recycling of polyester. These recycling options were compared to incineration (which was the option for conventional textile waste treatment in Sweden). From the results, it was concluded that incineration has the highest GWP and primary energy usage. The material reuse process has showcased the best performance amongst all the systems considered in this study, with the savings of 8 ton of carbon dioxide equivalents (CO_2-eq) and 164 GJ of primary energy per ton of textile waste.[18]

8.4.1 Challenges for recycling of textile products

Textile recycling is slowly becoming more popular and many products are being prepared that incorporate recycled fibres/materials. However, there are still challenges to be faced in the recycling of textile products.

- Developing a high-quality recycled material that can completely replace a virgin material is difficult. Developing an inferior-quality material with not-so-good mechanical and physical properties, which can therefore be used in producing low-quality products (such as fillings in mattresses or for upholstery product development), is relatively straightforward.[16]
- As a general rule, because recycling is usually a downcycling process, the quality of the recycled material is not on a par with the virgin equivalent.
- With a view to obtain a better quality recycled product, certain precautions must be taken when manufacturing the original textile product (at the first stage of the life cycle). For example, the mixing of different fibres or use of inferior-quality fibres should be avoided, the aim being to produce a product with a single fibre composition (e.g. 100% polyester).[3,16]
- Market potential for recycled products can be limited (this will be discussed in greater detail in Section 8.4.2).
- Technological problems and limitations of current recycling systems exist (both for mechanical and chemical recycling systems).
- Not all textile fibres have the potential to be recycled.
- Potential recycled fibres, such as carpets made out of nylon, do not have attractive market prices.[3]
- Improving the recovery rate by recycling most fibres is a challenging problem.
- The willingness and mindset of people have yet to reach a point where recycled fibres and recycled products gain wholehearted support. Still this remains the same, even though the concept of circular economy is gaining importance these days.

Detailed studies have yet to be reported on the recycling of textile materials in terms of open- and closed-loop recycling, or relating to different fibre materials, which behave differently when subject to recycling. In addition, difficulties associated with the processing of recycled materials alongside virgin materials, or separately, must be recorded. So far no such details are available to the scientific community. There are no readily accessible data that can be used in LCA relating to the recycling of different textile materials. The life cycle inventory analysis in relation to the recycling of various textile materials therefore handicaps textile researchers working in this area. Dedicated databases/datasets on the recycling of textile materials are much needed, so considerable research in these areas will be required for the textiles and clothing sector in the near future to fill such gaps.

8.4.2 Recyclability of textile fibres: the concept and quantification of recyclability potential index

As itemized in the earlier list, assessment of the recyclability of textile materials presents a considerable challenge. Recyclability is not a simple term and cannot be defined easily. No studies to date have defined or dealt with the assessment of recyclability of textile materials, with the exception of Muthu et al.[19] In this study, an initial attempt was made to explain the concept of recyclability of textile fibres by defining an index called the recyclability potential index (RPI). The RPI of various textile fibres was quantified using a scientific model developed by these researchers. Their RPI could not be defined by considering any one aspect, but rather took into account a combination of factors. The model they developed to quantify the RPI of textile fibres considered two important aspects: the environmental and the economic implications of recycling.

The environmental benefits considered in this conceptual model were

- conservation of essential resources such as energy and water used in the production of virgin materials;
- environmental impacts of producing virgin materials, including ecological and carbon footprint, as well as impact on human health;
- environmental impacts of landfilling the material instead of recycling.

The monetary benefits of recycled materials proposed by Villalba *et al.*[20] were considered in order to calculate the economic benefits of recycled materials. According to the model developed, RPI can be calculated by the following equation[21]:

$$\mathrm{RPI} = \Sigma \mathrm{EGI}_1 + \mathrm{EGI}_2 \qquad [8.1]$$

where EGI_1 is the environmental gain index and EGI_2 is the economic gain index. Furthermore,

$$\mathrm{EG}_1 = \Sigma \mathrm{X}_1 + \mathrm{X}_2 + \mathrm{X}_3 + \mathrm{X}_4$$

where X_1 denotes saving of potential resources, X_2 is the environmental impact caused by producing virgin fibres/materials, X_3 is the environmental impact due to landfilling and X_4 represents the environmental benefits gained from recycling as opposed to incineration. Also,

$$\mathrm{EG}_2 = \mathrm{x}_1/\mathrm{x}_2$$

where x_1 is the price of recycled fibre/material and x_2 is the price of virgin fibre/material.

The proposed concept and model of the RPI were applied to ten important textile fibres, namely, Nylon 6, Nylon 66, viscose, acrylic, polyester, wool, cotton, polypropylene and the polyethylenes low-density polyethylene (LDPE) and high-density polyethylene (HDPE).

In terms of environmental gains, the factor X_1 was determined by finding the energy and water requirements to produce 1 kg of the chosen virgin fibres. For the factor X_2, three major environmental impacts (namely, ecological and carbon footprint and the ecological damage in terms of human health) of producing 1 kg of the chosen virgin fibres were modelled with the aid of the software SimaPro version 7.2. The ecological footprint was modelled by Ecological Footprint V1.00, the carbon footprint was modelled using the IPCC (Intergovernmental Panel on Climate Change) 2007 GWP 100a method and the ecological damage was quantified by the Ecoindicator'99 method, in which only human health impacts were considered. For the factor X_3, the environmental impacts (namely, ecological and carbon footprints and the ecological damage in terms of human health) of disposing of 1 kg of the chosen textile fibres were also modelled with the aid of the LCA software SimaPro version 7.2. For the factor X_4, the environmental benefit of recycling versus incineration, suitable data for the chosen fibres were collected (i.e. energy conserved in kilowatt hours per ton in terms of recycling and energy generated in kilowatt hours per ton in terms of incineration).

To quantify the economic gain, the ratio of the price of recycled fibre to the price of virgin fibre of the same type and similar unit in the market must be acquired. Values for both recycled and virgin fibres obtained from the market were utilized to achieve this.

Collecting and quantifying these data for both indices generated a large number of values, with different units and under different headings. These had to be unified. To achieve this, a numerical scaling template was developed, consisting of five scales ranging from 1 to 5. These scales were also used to obtain the separate indices EGI_1 and EGI_2, and thereby to arrive at a value for RPI. The detailed methodology and results for individual factors of these ten fibres can be found in Ref. 18. As a result of this model, the RPIs of ten chosen fibres were quantified and ranked accordingly, as shown in Table 8.1.

The lower its RPI value, the better a fibre will be in terms of recyclability. Table 8.1 shows that polyester and polypropylene have the highest recyclability potential compared with the other fibres. They outperformed their counterparts and surpassed all the other fibres in both economic and environmental gains. Polypropylene scored better than polyester in terms of the environmental considerations used in this model. On the other hand, polyester was found to be better than polypropylene on economic grounds. Overall, these findings compensate each other so that these two fibres jointly occupy the top position in terms of RPI, followed by HDPE and LDPE.

Table 8.1 RPI of ten important textile fibres.[19]

Textile fibre	Environmental gain index	Economic gain index	RPI	Ranking in terms of recyclability
Nylon 6	30	2	32	6
Nylon 66	33	4	37	8
Viscose	29	4	33	7
Acrylic	24	3	27	4
Polyester	20	1	21	1
Cotton	25	4	29	5
Wool	27	5	32	6
PP	19	2	21	1
LDPE	21	2	23	3
HDPE	20	2	22	2

HDPE, high-density polyethylene; *LDPE*, low-density polyethylene; *PP*, polypropylene; *RPI*, recyclability potential index.

This is a relatively flexible model and can be applied to any fibre either using the same set of factors and data model or with a new set of factors and data model. Several factors are not taken into account owing to the lack of data: the technological challenges faced during materials sorting, different recycling policies/pressures exerted on various textile materials, availability of different fibres for recycling, rate and efficiency of recycling and quality of recycled materials compared to virgin materials. Furthermore, the scaling template developed and followed in this model is quite arbitrary; however, it could be further rationalized through an open discussion among various textile sectors and institutions.[19]

8.5 Incineration and landfilling of textile products

As discussed earlier, incineration with energy recovery is the next best option to recycling for waste management, as incineration without energy recovery clearly is not a desirable option. Incineration involves the burning of waste under controlled conditions and is responsible for various environmental and human health issues.

- residues emerging from incineration need further disposal and must also be tested.[21]
- concerns regarding transportation and its associated costs.[21]
- some undesirable by-products may arise following incineration.[21]
- there are issues relating to bottom and fly ashes produced by incinerators and their hazardous impacts.[22]
- emissions from incinerators may include heavy metals, dioxins and furans.[22]
- the process of incineration can create toxic products with associated undesirable impacts.[22]
- the various pollutants resulting from incineration have potentially detrimental effects on human health.[22]

These issues mean that in most countries, incineration is neither very common nor encouraged. Although textile products are not particularly harmful when incinerated, they do still pose certain threats to the environment and to human health.[2] Incineration of fibres produces a relatively lower amount of energy than that used to produce virgin fibres,[13] but the proportion of energy recovery differs from fibre to fibre and is decided by many factors. Data from some studies dealing with incineration of textile fibres[13,17,23] reveal that energy recovery from textile fibres ranges between 15% and 47%, considering a recovery rate of 60%. According to these studies, incineration of 1 ton of cotton and polyester can recover only 28% and 15%, respectively, of the total energy consumed during their production. These studies also showed that incineration of 1 ton of polypropylene could recover 34% of the energy used for virgin fibre production and 1 ton of viscose made in Asia and Austria could recover 15% and 47%, respectively, of the energy used for their virgin production. It was also noted that incinerating 1 ton of apparel could recover just 4% of the energy consumed in the vast production link between textiles and the clothing sector.[13,17,23] From these data, it is clear that incineration is certainly neither the best option in terms of saving resources nor the preferred option owing to the limitations outlined earlier.

Landfilling is the last method to consider and represents the least preferred option in the waste management hierarchy. Landfills are not generally encouraged by governments, NGOs or research scientists owing to their detrimental environmental effects (e.g. from landfill gas and leachates), their potential hazards to human health and the concerns over space availability. Indeed, in most countries the current landfills are full and there is simply no more space to accommodate new ones. Overall, in consideration of these factors, disposal at landfill sites remains the least preferred option for solid waste management, while other options such as reuse and recycling are encouraged. These days the option of landfilling is slowly diminishing due to the environmental and special considerations explained already. I am very happy to note that the recent LCA options (conceptually at least) demonstrate the end of life of a product by indicating only recycling rather than mentioning the end of life that includes all the options such as landfilling and incineration. What is importantly noted these days is that the option of landfilling is considerably seen to be reduced.

8.6 Biodegradation of textile products

Having discussed various solid waste management options and their implications, it is essential to consider biodegradation of products with specific reference to textiles and clothing materials. Biodegradation ensures safe disposal of a product and helps the product to complete its life cycle smoothly, without disturbing the environment. It is one of the main ways to divert waste from landfill disposal. Biodegradation refers to degradation by biotic means, defined as the chemical degradation of materials (i.e. polymers) aggravated by the action of microorganisms such as bacteria, algae and fungi.[24] Materials are thereby converted into water, biomass and carbon dioxide.[25]

There are two types of biodegradation, namely, aerobic and anaerobic. The presence or absence of oxygen during the process of biodegradation determines this distinction. If oxygen is present, aerobic biodegradation occurs and carbon dioxide is produced. If oxygen is absent, anaerobic degradation occurs and methane is produced instead of carbon dioxide.[25–28] The following basic equations explain the mechanisms of these processes:

Aerobic biodegradation:

$$\text{Polymer} + O_2 \rightarrow CO_2 + H_2O + \text{biomass} + \text{residue(s)}$$

Anaerobic biodegradation[24]:

$$\text{Polymer} \rightarrow CO_2 + CH_4 + H_2O + \text{biomass} + \text{residue(s)}$$

The extent of biodegradation a material undergoes depends largely on its polymer chemistry and the environment to which the material is exposed. A material's rate of biodegradation is influenced by many factors, such as

- the presence of microorganisms and oxygen availability,
- the temperature of the environment,
- the amount of water available for biodegradation,
- chemical factors (such as pH, electrolytes).[25,29]

There are various test methods and standards available to test the biodegradability of textile products, as issued by ASTM, ISO, OECD, EN (European) and AATCC. Biodegradation via soil burial test and the measurement of evolution of carbon dioxide from degraded samples are two of the most familiar methods followed for textile products. A number of studies have been conducted on textile product biodegradation, and some of these are discussed here.

Biodegradation studies on textile fibres and products have been conducted since 1949.[30] Microbiological deterioration of cellulose during the first 72 hours of attack was studied and recorded in 1950,[31] and the resistance to microbiological deterioration of resin-treated cellulosic fabrics was studied in 1954.[32] Between the 1960 and 1970s, many studies were conducted on the biodegradation of textile products and the microbial degradation of cellulosic textiles was studied in 1971.[33] Further studies were conducted in the 1990s, including degradability of cellulose acetate and cellophane in anaerobic bioreactors and in stimulated composting bioreactors in 1992,[34,35] anaerobic biodegradation of non-woven products in 1993[36]; and anaerobic degradation of cellulose in the same year.[37] The biological degradation of cellulose was studied in 1994.[38] Biodegradability of cotton, cellulose acetate fibres and their blends was studied in 1996.[39]

Biodegradation of various textile fibres when buried under a soil bed was studied and reported in 2003.[40] The soil burial test as per BS 6085, which is similar to AATCC 30, was followed for this study; flax, cotton, ramie, polypropylene and wool were the materials tested. The fibre samples were removed from the soil after 2, 4, 6, 8, 12, 16, 20 and 24 weeks and the biodegradation was studied both quantitatively and qualitatively. For the quantitative assessment, the weight loss of the fibres after burial for different periods was recorded; qualitatively, the fibres were examined under a scanning electron microscope. Results showed that among all the fibres considered, flax underwent the most rapid biodegradation process, followed by wool, cotton and ramie. Polypropylene was found to undergo very little biodegradation.[40]

A study conducted in 2004 analysed the biodegradability of cellulose fabrics made out of cotton, rayon, linen and acetate by using a soil burial test (AATCC 30), an activated sewage sludge test (ASTM D 5209-92) and an enzyme hydrolysis using *Trichoderma viride*. Surface changes after biodegradation were observed by optical microscopy. From the results of detailed tests and analysis, the extent of biodegradability was found to be higher in the case of rayon, followed by cotton and acetate in the decreasing order. Linen demonstrated an inconsistent behaviour: the amount of degradation it showed in the soil burial test was the greatest; however, it exhibited lower degradation than its counterparts in the other biodegradation tests.[41]

A study conducted in 2010 reported the biodegradability of 100% cotton knitted jersey fabrics with three levels of finishing treatments (scoured and bleached, softener added and resin added) and a polyester jersey fabric. This study was conducted in two stages: one under laboratory conditions and the other in a composting facility. The results obtained from each of these stages were compared. The standard ASTM D 5988-03 ('Standard test method for determination of aerobic biodegradation in soil of plastic materials or residual plastic materials after composting') was followed under laboratory conditions and included measurement of the amount of carbon dioxide produced. This was used alongside the quantification of the rate of biodegradation in natural soil by measuring the weight losses after biodegradation in enzyme solutions. The same set of fabrics was buried in a composting facility for a period of 3 months and the weight losses and fabric morphology after biodegradation were studied to quantify the biodegradability. Results showed that the polyester fabric demonstrated only slight initial degradation and afterwards the fabric remained intact under both laboratory and composting environments. On the other hand, the softener-treated cotton fabric was found to have an accelerated degradation rate, whereas the cotton fabric treated with resin was found to have a comparatively slow degradation rate. It was noted from this study that all three cotton samples demonstrated a noteworthy amount of degradation under the composting environment, confirming that the cotton samples were compostable.[42]

A study reported in 2011 involved the biodegradation study of cellulosic materials consisting of cotton, jute, linen, flax, wool and polyester fabrics, as well as one fabric with a blend of polyester and cotton. Biodegradation was tested under two conditions: by measuring the biodegradation of samples either in direct contact with the soil or not in direct contact with the soil. For the first condition, soil burial test as per ISO 11721-1: 2001 (Textiles – Determination of the resistance of cellulose-containing textiles to micro-organisms – Soil burial test – Part 1: Assessment of rot-retardant finishing) and ISO 11721:2003 (Textiles – Determination of the resistance of cellulose-containing textiles to micro-organisms – Soil burial test – Part 2: Identification of long-term resistance of a rot retardant finish) were followed for 3 months. For the second condition, all the chosen cellulosic samples were sewn into hydrophilic and hydrophobic bags, made out of polypropylene and polyethylene fabric at a ratio of 50/50. The fabric samples selected for this study were kept in the two bags and buried in soil as per ISO 11721-1:2001 and ISO 11721:2003 standards for 3 months. The second condition was intended to understand the differences in behaviour for all types of chosen cellulosic fabric samples in hydrophilic and hydrophobic conditions and to identify suitable conditions for biodegradation.[25] After every month, fabric samples were treated and measured for weight loss and their surface changes were studied by optical microscope, scanning electron microscope, Fourier transform infrared spectroscopy and thermogravimetric analysis.

The authors reported that it was difficult to judge the extent of biodegradation of samples when they were in contact with soil, as most of them were natural and cellulosic fibres. However, when samples were not in direct contact with soil, jute and linen were found to have biodegraded to a greater extent than cotton. Wool was found to be less biodegraded because penetration of microorganisms into wool is prevented by its scales and rigid molecular structure. Polyester had not biodegraded and, even in the flax-polyester blended fabric, only the cellulosic part was found to be affected by the microorganisms, not the synthetic polyester portion.[24]

Another study, reported in 2012, composed a soil burial test of various shopping bags used for groceries. This study was conducted as per the AATCC 30 standard for soil burial testing, and shopping bags made out of paper, woven cotton, HDPE,

LDPE, nonwoven polypropylene and nonwoven polyester were tested for 90 days in a purpose-built chamber under laboratory conditions. All the samples were tested for weight loss after 0, 30, 60 and 90 days and for loss/gain in tensile strength, strain and extension at 0 and 90 days, following suitable treatment and removal from the soil bed. Furthermore, any changes in appearance were measured using a microscope and the extent of biodegradation was assessed in terms of loss in both weight and tensile strength. From the results of the study, it was concluded that the cotton samples had undergone significant biodegradation in terms of both weight and strength loss. Polypropylene was found to perform better in terms of biodegradation when compared with its synthetic counterparts.[43]

Biodegradation is gaining importance in all fields and the development of biodegradable materials seems likely for every industrial sector, including textiles. Although many studies have been conducted relating to the biodegradation of textile materials, most have focused mainly on fabric stages, rather than on specific products. Some studies have investigated biodegradation of complete textile products,[43] but there is still a lack of relevant studies to cover the entire spectrum of textiles and clothing products, including apparels, technical textiles, nonwoven products, and so on. It still remains to create a dedicated information library compiling the biodegradation information for various textile fibres and blends. This should also be extended to various textile end products made out of those fibres and blends, with recommendations and suggestions to indicate the suitable conditions that must be followed for their biodegradation.

8.7 Summary

This chapter has focused mainly on the end-of-life management of textile products. The end-of-life phase is one of the most important phases in the life cycle of a product and determines the impacts of a product after its useful life ends. Once the decision has been made to discard any product, not just textiles, there are several potential destinations for it, namely, reuse, recycling, incineration or disposal at landfill. Each destination has its own benefits and limitations, and the associated implications for textiles and clothing products have been discussed in detail. A hierarchy of environmental benefits or gains can be defined, in which reuse is the favoured alternative, followed by recycling, then incineration, with disposal at landfill representing the worst option.

Reuse can be primary or secondary, and the associated environmental benefits were discussed in detail. Recycling falls into two types – open loop or closed loop – and each has its own pros and cons, which have been discussed in this chapter. The concept of recyclability is quite new to textiles, and textile recycling has its own challenges to be tackled, which were discussed at length. The basic concept and quantification of an RPI for various textile fibres in terms of environmental and economic gains were outlined and explained. Environmental implications of incineration and landfilling were also discussed. Each of these end-of-life destinations was discussed in relation to textile products, with suitable case studies. Finally, the concept of biodegradation and its types, along with the findings on biodegradation of various textile materials, was also described. Many gaps exist in the literature in terms of the end-of-life management of textiles, especially with respect to the recycling and biodegradation of textiles and clothing products, and these areas have been highlighted.

8.8 Sources of further information and advice

- Villanueva A, Delgado L, Luo Z, Eder P, Catarino AS, Litten D. *Study on the Selection of Waste Streams for End-of-Waste Assessment, Final Report.* Spain: European Commission, Joint Research Centre, Institute for Prospective Technological Studies; 2010.
- Allwood JM, Laursen SE, Russell SN, Malvido C, Bocken NMP. An approach to scenario analysis of the sustainability of an industrial sector applied to clothing and textiles in the UK. *Journal of Cleaner Production.* 2008;16:1234–1246.
- Baden S, Barber C. *The Impact of the Second-hand Clothing Trade on Developing Countries.* Oxfam; 2005
- Bartl A, Hackl A, Mihalyi B, Wistuba M, Marini I. Recycling of fiber materials. *Process Safety and Environmental Protection.* 2005;83(4):351–358.
- Chang Y, Chen H, Francis S. Market applications for recycled postconsumer fibers. *Family and Consumer Sciences Research Journal.* 1999;27:320–340.
- Chen H, Burns L. Environmental analysis of textile products. *Clothing and Textiles Research Journal.* 2006;24(3):248–261.
- Cliff R, Doig A, Finnveden G. The application of life cycle assessment to integrated waste management. Part 1 – methodology. *Transactions of the Institution of Chemical Engineers* 2000;78(B):279–287.
- Collins M, Gandy S. *Streamlined Life Cycle Assessment of Textile Recycling.* Oxford, UK: Salvation Army Trading Company Limited. Prepared by Environmental Resources Management Ltd; 2009.
- Desai AJ, Pandey SN. Microbial degradation of cellulose textiles. *Journal of Scientific and Industrial Research.* 1971;30:598–606.
- Fields S. *The Beneficial Nature of the Second-Hand Clothing Trade in Sub-Saharan Africa*; 2004.
- Finneveden G, et al. Solid waste treatment within the framework of life cycle assessment. *Journal of Cleaner Production.* 1995;3(4):189–199.
- Fletcher H. Disposable fashion: for sale, hardly worn, two million tonnes of cloths. *Times Online.* 2008. Available from: http://women.timesonline.co.uk/tol/life_and_style/women/fashion/article5332366.ece.
- Gulich B. Designing textile products that are easy to recycle. In: Wang Y, ed. *Recycling in Textiles.* 1st ed. Cambridge, UK: Woodhead Publishing Limited; 2006:25–37.
- Hansen KT. Dealing with used clothing: Salaula and the construction of identity in Zambia's third republic. *Public Culture.* 1994;6:503–523.
- Hansen K. Helping or hindering? Controversies around the international second-hand trading trade. *Anthropology Today.* 2004;20(4):3–9.
- Hansen K. Crafting appearances: the second hand clothing trade and dress practice in Zambia. In: Palmer A, Clark H, eds. *Old Clothes, New Looks: Second-Hand Fashion.* 1st ed. Oxford, UK: Berg Publishers; 2005:103–117.
- Hawley J. Digging for diamonds: a conceptual framework for understanding reclaimed textile products. *Clothing and Textiles Research Journal.* 2006;24(3):262–275.

- Hawley J. Economic impact of textile and clothing recycling. In: Hethorn J, Ulasewicz C, eds. *Sustainable Fashion: Why Now? A Conversation Exploring Issues, Practices, and Possibilities.* 1st ed. New York: Fairchild Books; 2008:207–232.
- Hawley J. Textile recycling as a system: a micro/macro analysis. *Journal of Family and Consumer Sciences*. 2000;93(5):35–40.
- Chen HL, Cluver B. Biodegradation and mildew resistance of naturally colored cottons. *Textile Research Journal*. 2010;80:2188.
- Limnander A. Please recycle. *Vogue*. 2000;290:297.
- Warnock M, Davis K, Wolf D, Gbur E. Soil burial effects on biodegradation and properties of three cellulosic fabrics. *AATCC Review*. January/February 2011.
- Collins M, Gandy S. *Streamlined Life Cycle Assessment of Textile Recycling. Final Report Prepared by Environmental Resources Management.* Salvation Army Trading Company Ltd; February 2002.
- Morley N, Slater S, Russel S, Tipper M, Ward GD. *Recycling of Low Grade Clothing Waste.* Oakdene Hollins Ltd, Salvation Army Trading Company Ltd, Nonwovens Innovation and Research Institute Ltd; 2006.
- Nakano Y. *Perception Towards Clothes with Recycled Content and Environmental Awareness: The Development of End Markets.* UK: Centre for Design Research, Northumbria University; 2006.
- Norris L. Cloth that lies: the secrets of recycling in India. In: Kuchler S, Miller D, eds. *Clothing as Material Culture.* Basingstoke, UK: Berg Publishers; 2005:83–106.
- Parfitt J. *Waste Not, Want Not.* Banbury, UK: Waste and Resources Action Plan; 2002.
- Swiss Academy for Development. *Second-Hand Clothing: Export, Social Compatibility and Social Acceptance. A Study on the Trade in Donated Clothes in Switzerland, Ghana and Tunisia.* Solothurn, Switzerland: SAD; 1997.
- Szostak-Kotowa J. Biodeterioration of textiles. *International Biodeterioration and Biodegradation*. 2004;53:165–170.
- Domina T, Koch K. The textile waste lifecycle. *Clothing and Textiles Research Journal.* 1997;15(2):96–102.
- Wang Y. *Recycling in Textiles.* Cambridge: Woodhead Publishing Limited, Woodhead Publishing Series in Textiles No. 50, ISBN 1 85573 952 6; March 2006.

8.9 References

1. European Council and Parliament. Directive 2008/98/EC of 19 November 2008 on waste and repealing certain directives. *Official Journal of European Union*. 2008;L312:3–30.
2. Palm D. *Improved Waste Management of Textiles, Project 9 Environmentally Improved Recycling*. IVL Report B1976. IVL Swedish Environmental Research Institute Ltd; April 2011.
3. Morley NJ, Bartlett C, McGill I. *Maximising Reuse and Recycling of UK Clothing and Textiles. A Report to the Department for Environment, Food and Rural Affairs*. Oakdene Hollins Ltd; 2009.
4. Ljunggren Söderman M, Palm D, Rydberg T. *Förebygga Avfall Med Kretsloppsparker. Report B1958*. Göteborg: IVL Swedish Environmental Research Institute Ltd; 2011.
5. Sundqvist J-O, Palm D. *Miljöpåverkan Från Avfall – Underlag För Avfallsprevention Och Förbättrad Avfallshantering. Report B1930*. IVL Swedish Environmental Research Institute Ltd; 2010.

6. McGill M. *Carbon Footprint Analysis of Textile Reuse and Recycling* [M.Sc. thesis]. Centre for Environmental Policy, Imperial College London, Faculty of Natural Sciences; September 2009.
7. Fisher K. *Impact of Energy from Waste and Recycling Policy on UK Greenhouse Gas Emissions*. Oxford, UK: Department for Environment, Food and Rural Affairs. Prepared by Environmental Resources Management Ltd; 2006.
8. Fisher K, Collins M, Aumônier S, Gregory B. *Carbon Balances and Energy Impacts of the Management of UK Wastes*. Oxford, UK: Department for Environment, Food and Rural Affairs. Prepared by Environmental Resources Management Ltd; 2006.
9. Muthu SS, Li Y, Hu JY, Mok PY. Carbon footprint of shopping (grocery) bags in China, Hong Kong and India. *Atmospheric Environment*. 2011;45:469−475.
10. Woolridge A, Ward B, Phillips P, Collins M, Gandy S. Life cycle assessment for reuse/recycling of donated waste textiles compared to use of virgin material: a UK energy saving perspective. *Resources, Conservation and Recycling*. 2005;46:94−103.
11. Farrant L, Olsen SI, Wangel A. Environmental benefits from reusing clothes. *International Journal of Life Cycle Assessment*. 2010;15:726−736.
12. Zamani B, Sandin G, Peters G. Life cycle assessment of clothing libraries: can collaborative consumption reduce the environmental impact of fast fashion? *Journal of Cleaner Production*. 2017;162:1368−1375.
13. Bartl A. *Fiber Recycling: Potential for Saving Energy and Resources*. Vienna, Austria: Vienna University of Technology − Institute of Chemical Engineering; 2008.
14. Allwood JM, Laursen SE, de Rodríguez CM, Bocken NMP. *Well Dressed?* University of Cambridge; 2006.
15. Muthu SS, Li Y, Hu JY, Ze L. Carbon footprint reduction in the textile process chain: recycling of textile materials. *Fibers and Polymers*. 2012;13(8):1065−1070. https://doi.org/10.1007/s12221-012-1065-0.
16. Zamani B. *Carbon Footprint and Energy Use of Textile Recycling Techniques, Case Study* [Master of Science thesis]. Sweden: Department of Chemical and Biological Engineering, Chalmers University of Technology; 2011.
17. Shen L, Patel M. Life cycle assessment of polysaccharide materials: a review. *Journal of Polymers and the Environment*. 2008;16:154−167.
18. Zamani B, Svanström M, Peters G, Rydberg T. A carbon footprint of textile recycling: a case study in Sweden. *Journal of Industrial Ecology*. 2014;19. https://doi.org/10.1111/jiec.12208.
19. Muthu SS, Li Y, Hu JY, Mok PY. Recyclability Potential Index (RPI): the concept and quantification of RPI for textile fibres. *Ecological Indicators*. 2012;18:58−62.
20. Villalba G, Segarra M, Fernández AI, Chimenos JM, Espiell F. A proposal for quantifying the recyclability of materials. *Resources, Conservation and Recycling*. 2002;37(1):39−53.
21. *Incineration, Waste Management Options*. US EPA Website. Available at: http://www.epa.gov/osw/homeland/options.htm#incineration.
22. Safar S. *Negative Impacts of Incineration-Based Waste-to-Energy Technology, Alternative Energy*; September 8, 2008. Available at: http://www.alternative-energy-news.info/negative-impacts-waste-to-energy/.
23. Koslowski HJ. *Dictionary of Man-Made Fibers*. Special ed. Duisburg, Germany: Sachtleben Chemie GmbH, International Business Press Publishers; 2000.
24. Leja K, Lewandowicz G. Polymer biodegradation and biodegradable polymers − a review. *Polish Journal of Environmental Studies*. 2010;19(2):255−266.
25. Arshad K, Mujahid M. *Biodegradation of Textile Materials* [Master's thesis for the Master in Textile Technology]. Sweden: The Swedish School of Textiles, University of Boras; 2011.

26. Kyrikou J, Briassoulis D. Biodegradation of agricultural plastic films: a critical review. *Journal of Polymers and the Environment*. 2007;15:125.
27. Grima S, Bellon-Maurel V, Feuilloley P, Silvestre F. Aerobic biodegradation of polymers in solid-state conditions: a review of environmental and physicochemical parameter settings in laboratory simulation. *Journal of Polymers and the Environment*. 2002;8:4.
28. Swift G. Requirements for biodegradable water-soluble polymers. *Polymer Degradation and Stability*. 1998;59:19.
29. Van der Zee M, Stoutjesdijk JH, Van der Heijden PAAW, De Wit D. Structure-biodegradation relationships of polymeric materials and effect of degree of oxidation of carbohydrate polymers. *Journal of Polymers and the Environment*. 1995;3(4):235–242.
30. Siu RGH, Darby RT, Burkholder PR, Barghoorn ES. Specificity of microbiological attack on cellulose derivatives. *Textile Research Journal*. 1949;19:484–488.
31. Abrams E. Microbiological deterioration of cellulose during the first 72 hours of attack. *Textile Research Journal*. 1950;20:71–86.
32. Cooke TF. Resistance to microbiological deterioration of resin-treated cellulosic fabrics. *Textile Research Journal*. 1954;24:197–209.
33. Desai AJ, Pandey SN. Microbial degradation of cellulosic textiles. *Journal of Scientific and Industrial Research*. 1971;30:598–606.
34. Gu J, Gada M, McCarthy S, Gross R. Degradability of cellulose acetate and cellophane in anaerobic bioreactors. *Polymer Material Science and Engineering*. 1992;67:230–231.
35. Gu J, Gada M, McCarthy S, Gross R, Eberiel D. Degradability of cellulose acetate and poly(lactide) in stimulated composting bioreactors. *Polymer Material Science and Engineering*. 1992;67:351–352.
36. Lindsay A, Woodings C. *Anaerobic Biodegradation of Non-woven Products*. UK: Courtaulds Research; 1993:78–80.
37. Buchanan CM, Gardner RM, Komarek RT. Aerobic biodegradation of cellulose acetate. *Journal of Applied Polymer Science*. 1993;47:1709–1719.
38. Beguin P, Aubert J. The biological degradation of cellulose. *FEMS Microbiology Reviews*. 1994;13:25–58.
39. Suh H, Duckett K, Bhat G. Biodegradable and tensile properties of cotton/cellulose acetate nonwovens. *Textile Research Journal*. 1996;66:230.
40. Ansari IA, East GC, Johnson DJ. Structure–property relationships in natural cellulosic fibres: Part IV: biodegradability of textile fibres. *Journal of the Textile Institute*. 2003;94(1): 16–36.
41. Park CH, Kang YK, Soon Im S. Biodegradability of cellulose fabrics. *Journal of Applied Polymer Science*. 2004;94:248–253.
42. Li L, Frey M, Kristie JB. Biodegradability study on cotton and polyester fabrics. *Journal of Engineered Fibers and Fabrics*. 2010;5(4):42–53.
43. Muthu SS, Li Y, Hu JY, Mok PY. A societal hot-button issue: biodegradation (soil burial test) studies of grocery shopping bags. *Energy, Education, Science and Technology*. 2012; 29(1):31–40.

Measuring the environmental impact of textiles in practice: calculating the product carbon footprint and life cycle assessment of particular textile products

9

9.1 Introduction

The previous chapters in this book have discussed the textiles and clothing supply chain, environmental regulations applicable to textile products, the concept of life cycle assessment (LCA) and its implications for the textiles and clothing supply chain, the different types of environmental impacts to be assessed and their assessment methodologies, the modelling of life cycle and product carbon footprint (PCF) assessments of textile products and, finally, the end-of-life management of textile products. Those chapters provide the necessary background for the specific case studies discussed in this chapter, which will focus on measuring the carbon footprint and various other environmental indicators of particular textile products using the LCA methodology. As mentioned in earlier chapters, the carbon footprint is one of the most important environmental indicators to be measured in LCA methodology and, in consequence, will be given special emphasis in this chapter.

There have been many important research studies focused on the various life cycle phases of textile products, but of all the phases, the consumer use phase is the most significant. Although the exact impact of the consumer use phase in the overall life cycle varies for different products (and many factors come into play here, such as the fibre type, care and maintenance needs and the geographical location in which the product is to be used), by and large the consumer use phase decides the entire fate of a product in terms of life cycle impact. Furthermore, even for the same product, the impact depends completely on consumer behaviour, which will vary greatly. This point has not been addressed in the previous chapters, and it will be dealt with here for various textile products. This chapter will also discuss the important product categories covering the entire textile supply chain, including cotton clothing, synthetic clothing, non-woven products and technical textile products, with examples in each category discussed in detail.

Assessing the Environmental Impact of Textiles and the Clothing Supply Chain
https://doi.org/10.1016/B978-0-12-819783-7.00009-0

9.2 Cotton clothing: life cycle assessment studies of T-shirts

Cotton T-shirts have been most commonly chosen by researchers for carbon footprint and LCA study. This chapter includes some of the important studies carried out on cotton T-shirts, measuring the carbon footprint and other environmental indicators.

9.2.1 Carbon footprint of a T-shirt manufactured in India, used and disposed of in the United Kingdom

A study by Continental Clothing was conducted in 2009 in three stages to cover the entire supply chain of a T-shirt, from cradle to grave, using PAS 2050:2008 (specification for the assessment of the life cycle greenhouse gas [GHG] emissions of goods and services). By choosing one garment (a typical women's short-sleeved T-shirt of small size, printed with one colour), this study aimed to quantify GHG emissions throughout its life cycle. This study also focused on locating the hot spots in terms of GHG emissions throughout the life cycle of the product. The T-shirt in this study was manufactured and dyed in India, the designs on the T-shirt were silk screen printed in Bristol and the T-shirts were worn, washed and disposed of in the United Kingdom. The sample was certified organic and was manufactured in a low-carbon environment. The study showed that 2.34 kg CO_2e was emitted for the selected product throughout its life cycle.[1]

The breakdown of this total figure into individual life cycles reveals that the 'use' phase was the most detrimental, responsible for 48% of total life cycle impacts (1.13 kg CO_2e), whereas all the other phases together created only 52% of the total life cycle impacts: 0.36 kg CO_2e from raw materials and manufacturing, 0.09 kg CO_2e from screen printing and 0.24, 0.19 and 0.43 kg CO_2e from transportation, packaging and retail, respectively. Disposal contributed a benefit/gain in terms of global warming (−0.1 kg CO_2e). During the consumer use phase the sample was assumed to be washed in an automatic washing machine 25 times during its lifetime, which, along with tumble drying and ironing, accounted for the 48% of total GHG emissions. Of the 'use' phase impacts, tumble drying was found to have the greatest emissions (45%), followed by ironing (31%) and automatic washing (24%). The results of this study further revealed that the avoidance of tumble drying and ironing could bring down the total GHG impacts by 37% or 0.857 kg CO_2e.[1]

Other key findings were also revealed from this study: around 90% reduction in manufacturing emissions can be achieved if renewable energy can be used, transportation by ocean freight has negligible contribution to GHG emissions compared with road transportation, minimum and/or reusable packaging is preferred and adding more colours to the design increases the GHG burden. Finally, garments should be given a cold wash, tumble drying should be avoided and ironing should be done only if necessary.[1]

9.2.2 Comparative life cycle assessment of a 100% Australian cotton T-shirt and a polyester T-shirt

This study was conducted in 2009 by the Institute for Sustainable Resources, at the Queensland University of Technology (QUT), Australia. The study analysed the GHG emissions, fossil fuel usage, ozone layer and mineral resources depletion impacts over the entire life cycle of both cotton and polyester (PET) T-shirts from cradle to grave, encompassing the following stages:

- cotton growth and ginning and manufacturing of PET (synthesis and extrusion);
- textile manufacturing processes;
- consumer use and maintenance phases of T-shirts including washing, drying and ironing;
- disposal.

From this study, it was found that 1 kg of cotton production emits 3.3 kg CO_2e, whereas 1 kg of PET emits 20 kg CO_2e. When it comes to overall results of the LCA, the cotton T-shirt was found to be better than the PET one in all the impact categories studied. The manufacturing phase was found to emit 12.5 kg CO_2e per kg of textile, where yarn production followed by wet processing were found to be the major parts, with 5.2 and 4.7 kg CO_2e per kg of textile produced, respectively. Also it was found from this study that the consumer phase was the most detrimental, contributing to 96% of the total life cycle impacts (277 kg CO_2e/T-shirt for 75 wash–dry operations over its lifetime).[2]

9.2.3 Carbon footprint of a 100% organic cotton T-shirt manufactured in India

This study was conducted in 2010 to analyse the carbon footprint of the supply chain for 100% organic cotton, raw material to retail sale. Raw material comes from Maharashtra, India, and the manufacturing phases of the T-shirts, covering spinning, steaming, knitting, dyeing, cutting and sewing, and packing, also occur in India. Transportation from the factory gate to the port of Mumbai and shipment to South African ports is followed by distribution to the retail clothing stores in South Africa. Retail sale, use and disposal phases were not a part of this study.[3,4]

The study revealed emissions of 1922.8 kg CO_2e per tonne of T-shirts, which works out to 5.455 kg CO_2e per T-shirt (4000 T-shirts/tonne). When the results were analysed the manufacturing phase was found to be the major part, with an emission of 4.77 kg CO_2e per T-shirt, followed by the raw material stage (farming and ginning) with an emission of 0.575 kg CO_2e per T-shirt. The contribution of storage and transportation to GHG emissions was found to be only 0.11 kg CO_2e per T-shirt. Manufacturing alone contributed 88% of the total carbon footprint, followed by the raw material stage (11%), with the remaining 1% contributed by storage and transportation.[3] The study recommended that manufacturers seek ways of reducing GHG emissions from the manufacturing phase, such as by utilizing renewable energy measures. It was indicated in the study that switching from grid electricity to renewable sources, such as biomass, water or wind power, and the like, could greatly lower GHG emissions in the manufacturing stage (from 4.77 kg CO_2e per T-shirt to 0.074 kg CO_2e per T-shirt). Similarly, measures were also recommended to reduce the GHG emissions from the raw material and storage and transportation phases.[3,4]

9.2.4 Study of life cycle inventory (quantification of energy use and four major pollutants) of a cotton T-shirt produced in India and consumed in Germany

This study was conducted in 2009, covering the following processes for the 'cradle to grave' of a cotton T-shirt:

- agricultural phase and ginning;
- textile processes including spinning, knitting, bleaching, dyeing and manufacturing;
- packaging;
- transportation in India, in Germany and intercontinental
- retailing operations in Germany;
- use phase (washing and drying);
- disposal (incineration/landfill).

Throughout the entire process, elements such as direct and indirect energy usage, use of chemicals and use of water were considered. The usage of electricity, detergent manufacturing and water treatment were included in assessing the use phase. The functional unit of the study was '100 days of a garment being worn' and the chosen T-shirt, weighing 250 g, was estimated to be washed once every two wears, i.e. 50 times total during its life. The life cycle inventory (LCI) analysis of this study primarily involved quantifying four important pollutants, namely, carbon dioxide (CO_2), nitrogen oxides (NO_x), particulate matter (PM) and sulphur dioxide (SO_2), as well as the primary non-renewable energy.[5]

From this study, a T-shirt in its entire life cycle, with the conditions assumed, used 240.11 MJ of energy, with drying and washing being responsible for the majority. Drying 50 times alone consumed 114.23 MJ, followed by washing 50 times, which consumed 62.16 MJ. Concerning the four major pollutants, in the lifetime of 100 wears, one T-shirt was found to emit 14 kg CO_2, 48.98 g NO_x, 32.70 g PM and 32.03 g SO_2. The use phase, involving washing and drying, was the major source of emissions of CO_2 and PM. Cotton fibre production, followed by drying and washing, was the major contributor of SO_2 emissions, while the major contributors of NO_x emissions were the cotton fibre production process followed by the dying process (wet textile processes). Many different use scenarios were analysed, and the following recommendations were made to reduce the use phase impact:

- more than 60% reduction in impacts can be achieved if a consumer opts for air drying instead of machine drying,
- 30% reduction of the washing phase emissions can be achieved at 40 °C compared with 60 °C,
- 20% reduction can be achieved by choosing A-rated appliances over C-rated ones (other parameters held constant).[5]

9.3 Cotton clothing: life cycle assessment studies of jeans

9.3.1 *Study by Levi Strauss & Co*

An LCA of jeans was conducted by Levi Strauss & Co. in 2006. A medium stone-washed product (produced from a shrink-to-fit fabric) produced for the American market in 2006 was selected for this cradle-to-grave assessment. All the processes covered in the life cycle of jeans are

- cotton cultivation and production,
- transportation of cotton to fabric production mills,
- fabric manufacturing processes,
- garment cut and sew processes,
- garment finishing processes,
- intermediate transportation,
- distribution and retailing,
- use phase,
- disposal phase (reuse, landfill and incineration).

Data were collected and compiled from the suppliers of Levi Strauss & Co. and, additionally, datasets from GaBi 4 software were used. The study was conducted as per ISO 14040 standards and focused on three major impacts, namely, climate change, energy use and water consumption. Functional unit details pertaining to this study were not clearly obtained from any of the respective references, even after repeated efforts spent in searching. It is assumed to be a unit of jeans. From the LCA results, it can be seen that the consumer use phase (58% contribution) creates the major impact as far as climate change is concerned, followed by the textile manufacturing stage from fibre to fabric (21%). The detailed results for individual phases are cotton − 1.7 kg CO_2e, fabric production − 6.6 kg CO_2e, cut/sew/finish − 3.0 kg CO_2e, logistics and retailing − 2.1 kg CO_2e, consumer use − 18.6 kg CO_2e and end of life − 0.5 kg CO_2e.

As expected, the trend in terms of energy use impacts was similar. The results of individual phases in terms of energy use are cotton − 17.7 MJ, fabric production − 84.9 MJ, cut/sew/finish − 40.8 MJ, logistics and retailing − 29.8 MJ, consumer use − 226.6 MJ and end of life − 0.4 MJ. With respect to water consumption impacts, cotton production (49% contribution) surpasses the consumer use phase (45%). Cotton production was found to consume 1704 litres of water, and fabric production and cut/sew/finish consumed 72.1 and 110.8 litres of water, respectively. Logistics and retailing operations consumed 18.1 litres of water, and consumer use and end-of-life phases consumed 1575.2 and 0.4 litres of water, respectively.[6]

As the consumer use phase was found to be the key to reduce the major impacts studied, this study recommended that consumers consider reducing the number of washes in the lifetime of the jeans. The study results confirmed that switching to washing the jeans once every 2 weeks or once a month, instead of once each week, could reduce, respectively, climate change impact by 32% and 48%, energy use by 20% and 40% and water consumption by 23% and 35%.[6]

9.3.2 *Environmental product declaration of jeans*

An environmental assessment study of a pair of jeans was conducted by the Bio Intelligence Service in 2006 for the Department of Eco-Design and Sustainable Development of the French Environment & Energy Management Agency – ADEME, and the details of this study were published in the form of an Environmental Product Declaration (EPD) for jeans. This study was conducted as per ISO 14040 standards, and the EPD was compliant with ISO 14025 standards (environmental labelling of type III). The functional unit chosen for this study was 'wear jeans for 1 day', which means the life cycle impacts pertaining to wearing a selected pair of jeans for a day were quantified.[7]

The pair of jeans selected for this study was made from slightly washed out blue denim material, weighed 666 g, including 600 g of denim material, 37.5 g of lining fabric, 10.4 g of double yarn, 3.6 g of rivets (for a total of six rivets) and 14 g of buttons (for a total of four buttons). As indicated in the defined functional unit, the selected jeans were assumed to be worn 1 day per week for the 4-year life span, and they were assumed to be washed after wearing three times, in a class C type of washing machine at 40 °C. Disposal options for the selected jeans were throwing away with household waste (50%) or given or sold to a second user, who will use them for a further 4 years.[7]

Data pertaining to the manufacturing processes were obtained from a company based in Tunisia, which produced jeans for a French brand, and the use and disposal scenarios were modelled as per the French conditions. Other details pertaining to the assumptions, boundaries, data gaps and collection of data for other phases can be found in the report.[7] Results of the study for the chosen eight environmental impact indicators per functional unit selected for this study are as follows: primary energy consumption – 1.49 MJ; climate change – 44 g CO_2 equivalent (eq); ozone layer depletion – 0.000004 g CFC-11 eq; human toxicity – 31 g 1,4-dichlorobenzene eq; aquatic ecotoxicity (fresh water) – 51 g 1,4-dichlorobenzene eq; eutrophication – 0.04 g PO_4^{2-} eq; water consumption – 0.02 m^3; and production of household waste – 6 g. In addition, detailed recommendations to reduce the impacts arising from the selected product were discussed.[7]

9.4 Woollen clothing: life cycle assessment study of a sweater

This case study revolves around a PCF study conducted on a simple wool sweater (without buttons, laces, zippers or other accessories), which was produced by standard production processes, including yarn production, weaving processes and finishing treatments. LCA methodology was followed to quantify the carbon footprint in accordance with ISO 1404X standards by the IPCC 2007 method (100-year time frame was chosen) with the aid of SimaPro LCA software. The Ecoinvent database was referred to for the production phases of the woollen sweater. Boundaries of assessment included processes from sheep breeding at the South African farms, wool scouring, dyeing, spinning, knitting, transportation to the distribution centres and then to the retail stores, packaging, washing and final disposal. An average sweater (weighing 264.85 g without accessories) was considered as a functional unit, although more accurately; this is one of the primary issues of this study. The chosen sweater was made from 100% merino wool, in four colours. LCI was collected from both primary and secondary data sources, and the details can be found in the study.[8]

An average life of 5 years was assumed for the chosen product for this study. For the use phase, it was assumed that the product was washed 15 times per year at 30 °C. For each wash, it was estimated that 10 litres of water and 130 mL of chemicals (detergent and conditioner) were required. The end-of-life phase was modelled as 49% disposed, 49% burnt and 2% reused. The results showed that over its total life cycle the garment emitted 1947 kg CO_2. The distribution phase caused the greatest impact (0.470 kg CO_2), followed by the finished garment processing (0.384 kg CO_2). The sheep breeding and the washing phases were found to be the next largest contributors to the carbon footprint, with 0.376 and 0.280 kg CO_2, respectively. The transportation and spinning processes emitted 0.218 and 0.142 kg CO_2, respectively. Other processes, such as scouring of the wool and dyeing of the flocked wool, contributed little to the final result, and final disposal was found to have nil GHG emissions. An uncertainty assessment and a sensitivity study were also conducted, and the details can be found in reference.[8]

9.5 Synthetic clothing: life cycle assessment studies of polyester garments

9.5.1 Life cycle assessment of a polyester blouse

This is a study conducted in 1995 to examine the LCI of a PET blouse. The study was confined to LCI, although it was described as LCA. This was a cradle-to-grave study performed by Franklin Associates, Ltd for the American Fiber Manufacturers Association (AFMA) on a women's blouse made from 100% PET knit fabric. The study quantified the energy requirements, atmospheric emissions, waterborne wastes and solid wastes at both industrial and consumer stages. The sample chosen was a short-sleeved, pullover women's blouse made from 100% knitted PET filament weighing 0.12 lbs, without any accessories such as buttons or zippers, and the study included the primary and secondary packaging associated with the product. The study encompassed the following processes:

- extraction of raw materials and further processing,
- fabric production processes,
- apparel production processes,
- transportation (at all stages),
- blouse laundering (machine washing and drying),
- disposal.[9]

Results were quantified for the functional unit of 1 million wears. The raw material requirements to equate to 1 million wears were quantified by assuming 40 wears as an average life span for a blouse. Load size of washing was assumed to be 20 blouses, and the frequency of washing was assumed to be once after every two wears. From the results, it was shown that approximately 82% of the total energy used was during the consumer use phase, where two-thirds of the total energy was needed for washing and the remaining one-third for drying. Resin manufacture and fabric production were the two major areas of production and consumed the greatest amount of energy. In terms of solid wastes, 66% of the total waste was created during the consumer use phase; blouse disposal contributed a further 24% to the total and the remaining 10% was attributed to the manufacturing process. As far as the atmospheric emissions are considered, airborne and waterborne emissions were quantified for 1 million wears. Most of the air emissions arose from electricity production demanded by the laundering process. Similarly, wastewater from the laundering process was the major contributor to waterborne emissions quantified in this study, apart from the releases generated during the burning of fossil fuels for energy generation.[9]

9.5.2 Study of the life cycle inventory (quantification of energy use and four major pollutants) of a polyester jacket produced in China and consumed in Germany

This study, conducted in 2009, revolved around the development of the LCI of a PET jacket, produced in China and consumed in Germany. This included the following aspects for assessment:

- resin production;
- spinning;
- knitting;
- apparel confection;
- transport in China, in Germany and intercontinental;
- use phase;
- disposal – landfill and incineration.

All the aspects of resin production, including the production of base ingredients and aspects such as direct and indirect energy use, packaging and water use, were considered in the production processes. Processes pertaining to electricity generation, detergent manufacturing and water treatment corresponding to washing and drying operations in the use phase were also included. The functional unit defined in this study was, '100 days of a polyester jacket, weighing 0.5 kg being worn, every 2 days, for two seasons'. It was assumed that the selected jacket was washed three times a season or six times during its entire lifetime. This study quantified four important pollutants, namely, carbon dioxide (CO_2), nitrogen oxides (NO_x), PM and sulphur dioxide (SO_2), as well as the primary non-renewable energy, in the entire life cycle of the selected PET jacket.[5]

The study showed that, for the assumed functional unit, over its total life cycle, the PET jacket consumed 156.39 MJ of energy. The production phase in China was the major element, consuming 71% of the total energy (resin – 26.82 MJ, yarn production – 29.23, fabric production – 52.92 MJ), followed by sale and use in Germany, which accounted for 28% of the total energy. Washing the garment six times in its entire life span consumed 14.92 MJ, and the equivalent drying consumed 27.41 MJ. In the total life cycle the jacket emitted 10.52 kg CO_2, 35.97 g NO_x, 21.50 g PM and 45.00 g SO_2. As with the energy use, the production phase in China was the major contributor for three of the four major pollutants studied; 72% of the CO_2 was emitted in the production phase, followed by 27% in sale and use in Germany; 88%, 85% and 92% of NO_x, PM and SO_2, respectively, were emitted during the production phase, followed by 8%, 15% and 5% of NO_x, PM and SO_2, respectively, during the sale and use phase in Germany.[5]

In contrast to the cotton T-shirt discussed in Section 9.2.4, and in other case studies, the use phase was not the major contributor to energy use and major emissions in the case of the PET jacket, where the production phase was found to be responsible for the greatest energy consumption and emission of pollutants.

9.6 Linen textiles: life cycle assessment of a linen shirt

The eco-profile of a linen shirt and a comparative study of linen and cotton shirts was conducted and published in 2008. This study was done by the Bio Intelligence Service for Masters of Linen. Twelve indicators were studied, but the results of only five indicators were presented in the eco-profile: primary energy consumption, water consumption, global warming potential, eutrophication and freshwater aquatic ecotoxicity potential. This cradle-to-grave study defined the functional unit as 'wearing a shirt for a day'. This study compared the environmental impacts of a linen shirt (cultivated in France, manufactured in China, transported to France for use and disposal) and a cotton shirt (cotton cultivated and the shirt manufactured in China and then transported to France for use and disposal). The chosen shirts of linen and cotton were of the same weight, 253 g, including 240 g of linen or cotton fabric, 10 g of PET yarn and 3 g of plastic buttons. It was assumed that both these shirts were worn on average 60 times before disposal by the primary user. The disposal scenario was assumed to be 30% disposal with the household waste and 70% reuse. With this scenario, the average total lifetime of a shirt worked out to 100 times of wearing. As far as the use phase was concerned, both types of shirt were assumed to be washed at 40 °C and ironed after each use, with different ironing periods (the linen shirt needed to be ironed at 9 minutes, whereas the cotton shirt required only 7 minutes).[10]

From the results of the LCA, it was revealed that wearing a linen shirt for 1 day consumed 6 MJ of energy, consumed 6.4 litres of water, emitted 130 g of CO_2 and is responsible for 0.10 g eq phosphate units of eutrophication and 11 g 1,4-dichlorobenzene eq of freshwater aquatic ecotoxicity potential. Results further revealed that the use phase was found to be the major factor in energy and water consumption and freshwater aquatic ecotoxicity potential, whereas the manufacturing phase was found to be the major contributor to global warming and eutrophication.

On comparative grounds, linen scored over cotton in terms of water consumption, eutrophication and freshwater aquatic ecotoxicity potential, but the position was reversed for the other two indicators studied.[10]

9.7 Technical textiles: life cycle assessment studies of medical textile products

Technical textiles is a vast topic, covering geo textiles, agro textiles and medical textiles. In this section, some of the case studies earmarked for LCA and PCF studies pertaining to medical textile products are discussed.

9.7.1 Whole life cycle inventory of medical gowns

This study focused on utilizing LCI to compare the environmental impacts of reusable and disposable biocidal healthcare garments. The functional unit chosen was aimed at comparing 75 disposable patient gowns made from polypropylene fabric, with reusable gowns made from 55% cotton and 45% PET. Environmental impacts, namely, the depletion of resources (water and non-water), energy use and air (CO_2), water and solid emissions, were quantified in this study.[11]

The results of the study showed that reusable gowns use significantly less resources than disposable gowns. When use is included, reusable gowns use more water resources, but less other resources. A similar trend was observed for energy use and emissions, i.e. disposable gowns used less in manufacturing, but more when usage was taken into account. Only in the case of water-based emissions alone did reusable gowns generate greater impact than the disposables. It was found that the reusable gown production consumed nine times more energy and seven times more resources than the disposable gowns, but the environmental impact of a reusable gown, laundered for 75 times, was less than the impact created by 75 disposable gowns, except for water use and water emissions.[11,12]

9.7.2 Life cycle assessment of surgical scrub suits

A study was conducted to quantify and compare the whole-life environmental impact of a single-use (disposable, made from polypropylene) and a multi-use surgical scrub suit (reusable up to 100 times, made from cotton), adopting ISO 14040 standards. A single use of a scrub suit of an average size, which fulfils the requirements of the European standard EN 13795, was chosen as the functional unit of the study. Impact categories selected for this study were climate change, eutrophication, PM formation, ozone depletion, ocean acidification, freshwater use, agricultural land occupation and total energy use, and the details of assumptions, scope, boundaries and allocation can be found in the study. The results showed that the disposable scrubs created greater impact in all the categories studied, compared with the reusable ones.[13]

Fabric manufacture and garment production were found to be the major contributors to the cumulative environmental impact of the disposable scrubs, with raw material production and fossil fuel usage for energy and transportation being the major factors. Fabric production and the final use phase were found to be the major contributors in the life of the reusable scrubs. Full details of the results can be found in Ref. 13, along with details of the sensitivity study.

If the reusable scrubs were not reused, but thrown away after a single use, then their impacts would be greater than those of the disposable ones; it is through the reuse that the impact in different categories is distributed and the total impact is reduced substantially. This is the case for all reusable items, not just the scrubs in this case.

9.8 Non-wovens: life cycle assessment studies of nappies (diapers)

Another textile category that needs to be considered via case studies is non-wovens, due to their widespread applications and current importance in the market. Non-wovens are used in many different categories of products, but the diaper is a very important product and is discussed in the following section.

9.8.1 Life cycle assessment of reusable and disposable nappies

There are many studies conducted on nappies (diapers), especially comparative studies on disposable and reusable ones. Readers are recommended a study by the author on the LCA studies on nappies, which details the crucial aspects addressed.[14]

Two studies conducted by the UK Environment Agency on the LCA of disposable and reusable nappies in the United Kingdom are discussed. The first study published in May 2005 was aimed at assessing the life cycle environmental impacts of disposable and reusable nappies in the United Kingdom during 2001–02. In this study, three different nappy types (disposable nappies, home-laundered flat cloth nappies and commercially laundered prefolded cloth nappies delivered to the home) were assessed for whole-life impact. This study quantified the following impact categories of the selected nappies for the functional unit defined for the study as 'the use of nappies during the first 2.5 years of a child's life, in the United Kingdom, for the period 2001–02':

- global warming,
- ozone depletion,
- photo-oxidant formation,
- depletion of abiotic resources,
- eutrophication,
- acidification,
- human toxicity,
- freshwater aquatic toxicity,
- terrestrial toxicity.

This study, conducted to ISO 14040 standards, excluded the environmental impacts associated with land occupation and use, and the systems evaluated were assumed to be steady state. The results showed that there was no significant difference between the three nappy systems studied with respect to life cycle environmental impact, and none of the systems were found to be less/more environmental friendly. Results showed that the disposable nappy system created greater environmental impact in the raw material production and nappy component manufacturing process. However, fuel and electricity consumption in the laundering process were the major sources of environmental impact for the home-laundered and commercially laundered nappies. It was also found that waste management did not contribute much to the overall impact of all the three nappy types studied. Detailed inventory analysis, impact assessment results and a sensitivity study can be found from the detailed report of this study.[15]

The UK Environment Agency conducted a revision of the above-mentioned study in 2008, with the aim of updating aspects of the marketplace between 2002/03 and 2005/06. The scope of this study was consistent with the previous study: product types, assumptions and limitations, and the functional unit were similar. New and important conclusions from this updated study are listed in the following:

- The manufacturing processes of disposable nappies has greater impact in the United Kingdom than their disposal by landfill.
- The impact of reusable nappies is entirely dependent on the way they are laundered.
- Contrary to disposable nappies, consumer behaviour after purchase determines the impact of reusable nappies.
- The impacts from cloth nappies can be reduced by using line drying as much as possible (and tumble drying as little as possible), using more energy efficient appliances, washing at temperatures below 60 °C and with fuller loads and reusing nappies on other children.[16]

9.8.2 Carbon footprint and eco-footprint of adult incontinence products

This study by the author deals with the modelling of the carbon and ecological footprints of adult incontinence products. A new type of diaper (eco-diaper) and a new type of reusable pant (eco-pant) were developed by one of the research teams of the Institute of Textiles and Clothing of The Hong Kong Polytechnic University, and these products were compared to commercial diapers and pants in terms of carbon and ecological footprints. The new types of eco-diapers and eco-pants are equipped with new types of layers, which enhance the comfort of the users, and the pants were mainly developed to be reusable and so tackle the waste management issue. Whole-life carbon footprint and eco-footprint of the four selected products were modelled using the IPCC 2001 method (100 years) and the eco-footprint method (taken from Ecoinvent 2.0) of SimaPro v.7.3.2 LCA software.[17]

Both the diaper types are disposable and the outer layer of both is made from polyethylene film, with the middle layer consisting of an absorbent layer of cellulose pulp and superabsorbent polymer (SAP). The inner layer of the commercial diaper was made from polypropylene non-woven material, whereas the newly developed one was made from 100% cotton non-woven fabric with moisture management treatment. Commercial disposable pants consisted of a polyethylene outer shell, a middle absorbent layer of cellulose pulp and SAP and an inner wicking layer made from polypropylene (non-woven). The outer layer of the new type of reusable pants was made out of 100% knitted cotton mesh and the inner layer was made of waterproof breathable cotton fabric and disposable inner pads (inserts) with an inner layer of cotton non-woven fabric with moisture management treatment. The middle layer – the inserts – was made from cellulose pulp and SAP, with an outer layer of breathable paper.[17]

The functional unit was defined as one product unit of each type of incontinence product. The study excluded the transportation phase of all four products from the manufacturing point to consumers, the excreta and the use phase of these products. Cradle-to-gate stage data were obtained from the Ecoinvent database. For the end-of-life phase, commercial and new diapers, commercial pants and inserts from reusable pants were assumed to be landfilled completely after the first use, and the reusable pants were assumed to be 90% reused and 10% landfilled.

The LCA results showed that the new reusable pants had lowest carbon footprint and eco-footprint compared with disposable ones, whereas the new type of diaper's footprint was slightly greater than that of the commercial one. From the detailed analysis of hot spots and process contributions, it was found that the production and disposal of SAP, the generation of electricity for polypropylene production and the production processes of cotton were the principal contributors to the footprint. It was also concluded that if the new type of reusable pants can be used without disposable inserts, they will significantly reduce the carbon footprint and eco-footprint.[17]

9.9 Recent life cycle assessment case studies

9.9.1 Life cycle assessment benchmarking study on five types of textiles

The goal of this study, which was published in 2014, was to provide an insight into the environmental burden of textiles made of the five base materials, namely, cotton, PET, nylon, acryl and elastane.

This study was aimed to answer the framed research question to find out the base material out of the selected five materials and which life cycle stage (cradle to gate as well as cradle to grave) has the biggest impact on the environment. LCI data for this study were collected from the literature, namely, LCA databases and emission registration database of the Dutch government, as well as from communications with both manufacturing companies of production equipment and textile companies involved in this study. The main focus of the study was on the production of textiles (cradle to gate); some data on the use phase (washing by the user) and the end-of-life phase were also provided. The choice of the declared unit (functional unit) was '1 kg of (greige) textile'.[18]

The output of the calculations was presented in four single indicators:

- eco-costs 2012 (a prevention-based indicator),
- CO_2 equivalent (carbon footprint),
- cumulative energy demand (CED),
- ReCiPe (a damage-based indicator)

As the purpose of this study was to provide designers with environmental information, the output of the calculations was not presented in the form of a set of midpoints, but in the form of single indicators (the indicators chosen were mentioned before). A single indicator in LCI analysis is one single score to express the result of the cumulative inventory list in one indicator, either at the midpoint or at the endpoint level.

Eco-costs is a measure to express the amount of environmental burden of a product on the basis of prevention of that burden, and these are the costs that should be made to reduce the environmental pollution and materials depletion in our world to a level, which is in line with the carrying capacity of our Earth.[19,20]

From the data analysis and interpretation, it was evident that the environmental burden is a function of not only the base materials (cotton, PET, nylon, acryl and elastane) but also the thickness of the yarn (for this research, the range of 50–500 dtex was examined). The cradle-to-grave analysis from raw material extraction to discarded textile demonstrated that textiles made out of acryl and PET have the least impact on the environment, followed by elastane, nylon and cotton. Cotton represented the highest environmental burden in all four single indicators (CO_2 equivalent, CED, eco-costs 2012 and ReCiPe) in this study. For cotton fabric less than 150 dtex, weaving and spinning were found to have the highest cradle-to-gate impact. For polymer fibers, the impact of spinning was found to be comparatively low; however, weaving had the highest impact in the production of textile less than 70 dtex. Knitting had a factor of 20 lower impact than weaving for all fibers, so knitting is mentioned in this study as a better solution than weaving from an environmental point of view. The total environmental burden over the complete life cycle was found to be reduced at a higher yarn size owing to the decrease in energy use (per kilogram textile) for the spinning and weaving processes of thicker yarns. As a consequence, the impact of weaving (and likewise of knitting) became less important and cotton and nylon change places in the ranking (breakeven point was around 100 dtex).

Another important and interesting finding from this study is that the use phase (the washing and drying of laundry) was found to have less relative impact than it is suggested in the classical literature. The authors of this study proposed that the environmental burden of spinning, weaving and knitting is a function of one/yarn size. The impact of spinning and weaving is relatively high (for yarn thicknesses of less than 100 dtex), and from the environmental point of view, knitting was found to be better than weaving (it is quite obvious though).[18]

Also this study has clearly mentioned that LCA on textiles can only be accurate when the yarn thickness is specified, and even in case the functional unit also indicates the fabric per square metre, the density must be known.[18]

9.9.2 Life cycle assessment of cotton T-shirts in China

Published in 2015, the goal of this study was to assess the environmental impacts and to identify the environmental hot spots, along with finding out improvement opportunities during the life cycle of 100% cotton T-shirts produced in China. The declared unit for this study was a piece of 100% cotton short-sleeved T-shirt (a 100% cotton knitted dyed short-sleeved T-shirt with density of 200 g/m^2 and mass of 0.153 kg).

The life cycle phases included in this study are cotton cultivation and cotton fiber production, including direct and indirect energy inputs, irrigation, commercial fertilizers and pesticides, and ginning (separation of seed and fiber); textiles manufacturing includes spinning, knitting, dyeing (pre-treatment, dyeing, post-treatment and finishing) and making up (cutting, sewing and package), with direct and indirect energy use, chemicals, and water use; and distribution, consumer use (wearing, washing, drying and ironing, taking electricity, detergent and wastewater emission into account) and disposal through landfill. Human labour, construction of capital equipment and maintenance and operation of support equipment were excluded from the system boundary.

The production data for this study were obtained from field investigations of representative mills in China. The use phase data were mainly from 924 questionnaires of Chinese residents. The secondary data were collected from databases, literatures and authoritative statistical data. The potential environmental impacts were evaluated using the CML 2001 and USEtox methodologies built into the GaBi version 6.0 LCA software.[21]

Life cycle impact assessment (LCIA) results of this study for the assumed functional unit based on CML 2001 and USEtox method are as follows: abiotic depletion potential – elements (ADP-e), 9.74×10^{-6} kg Sb-equiv.; ADP fossil (ADP-f), 58.1 MJ; acidification potential (AP), 0.0535 kg SO_2-equiv.; eutrophication potential (EP), 0.0186 kg P-equiv.; global warming potential (GWP), 6.05 kg CO_2-equiv.; photochemical ozone creation potential (POCP), 0.00316 kg ethene-equiv.; water use (WU), 1770 kg; ecotoxicity potential (ECP), 17.3 CTUeco; human toxicity potential – cancer (HTPC), 3.08×10^{-7} USEtox CTUh; and human toxicity potential – non-cancer (HTPNC), 9.46×10^{-7} USEtox CTUh.[21]

As far as ADPs (ADP-e and ADP-f) are concerned, the largest portion of the ADP (elements) burden was in the dyeing stage (58.04%) and the second contributor remained in the upstream cotton cultivation (28.66%). In terms of AP, the use stage (12.02%), dyeing stage (28.53%), cotton cultivation (16.71%) and making-up process (33.79%) covered the most AP. The use phase accounted for nearly half of the total EP of T-shirt, followed by the cotton cultivation stage, which contributed 25.76% to the total EP in the entire life cycle of a T-shirt selected in this study. The largest portion of GWP burden (carbon footprint) in the life cycle of one T-shirt was found to be from the dyeing process (34.79%) (mainly contributed by direct CO_2 emissions when hard coal is burned on-site to produce steam) and the second contributor to GWP was from the making-up stage (31.96%) (chiefly from the indirect CO_2 emissions because of the on-site electricity consumption). POCP is mainly attributed to direct NO_x emissions from coal burning in the dyeing stage and indirect SO_2 emissions from making-up and use stages because of electricity consumption. ECP was mainly from the cotton cultivation stage, which was accounting for 82.9% of the total ECP. The HTPC indicator was mainly arising from the dyeing process, which contributed to 58.9% to the life cycle burden of one T-shirt. The cotton cultivation, dyeing, making-up and use stages contributed 27.9%, 18.1%, 26.9% and 14.6%, respectively, to the total of HTPNC. As expected, cotton cultivation contributed nearly 80% of the life cycle water use of T-shirts, followed by the use stage and dyeing stage.[21]

The results of the study indicated that cotton cultivation, dyeing, making-up and use phases were the main contributors to the environmental impacts. In particular, fertilizer, pesticide and water use in cotton cultivation; coal, dyes and auxiliaries use in dyeing; electricity use in making up; detergent and water use in washing; and electricity use in spinning were the major hot spots based on the LCIA results. The use-phase scenario analysis has shown that compared with machine washing, electric drying and ironing shared the majority of electricity consumption. Energy consumption, chemical use and water use were found to be the main contributors to most impact categories analysed in this study.[21]

9.10 Summary

This chapter was principally aimed at highlighting the different aspects of measuring the environmental impact of textile products in practice through case studies of LCA and PCF of textiles and clothing products, including cotton clothing, synthetic clothing, woollen clothing, linen textiles, technical textile products and non-woven products. Only important case studies in each product/product sector were discussed: a cotton T-shirt, a pair of jeans, a PET blouse, a PET jacket, a linen shirt, a woollen sweater, a medical gown and various nappies (diapers). The details of each case study, such as the goal and scope, functional unit assumed, inventory details, impact assessment details, major results and conclusions, were highlighted, and some of the important issues found were indicated. Most of the case studies on clothing showed that the consumer use phase is the chief contributor to the ecological impact. The exception was the PET jacket, which does not require frequent washing and drying; it is critical to focus primarily on the use phase to reduce the environmental impacts of textiles and clothing products.

Of course, there are still key challenges in modelling the LCA and PCF of any product, including textiles and clothing, and it is important to highlight these challenges for the benefit of readers and researchers/students. Unfortunately, overall, this issue has not been addressed so far by previous studies. Other challenges and issues will be discussed in detail in the following chapter.

9.11 Sources of further information and advice

- Brent AC. A life cycle impact assessment procedure with resource groups as areas of protection. *The International Journal of Life Cycle Assessment.* 2004;9(3):172–179.
- Chen H, Burns DL. Environmental analysis of textile products. *Clothing and Textile Research Journal.* 2006;24 (3):248–261.

- Crews PC, Rich W, Niemeyer S. A summary of environmental legislation targeting disposable diapers and review of related literature. *Journal of Polymers and the Environment.* 1994;2(1):39–48.
- Fletcher K. *Sustainable Fashion and Clothing: Design Journey.* London, UK: Earthscan; 2008.
- Franklin Associates. *Energy and Environmental Profile Analysis of Children's Single Use and Cloth Diapers* (revised report). Prairie Village, Kansas: Franklin Associates; 1992.
- Guinée JB. (final ed), Gorrée M, Heijungs R, Huppes G, Kleijn R, *et al. Handbook on Life Cycle Assessment – Operational Guide to the ISO Standards.* Dordrecht: Kluwer Academic Publishers; 2004.
- IPCC. Summary for policymakers. In: Solomon S, *et al.*, eds. *Climate Change 2007: The Physical Science Basis.* Contribution of working group I to the fourth assessment report of the Intergovernmental Panel on Climate Change. Cambridge: Cambridge University Press; 2007.
- ISO 14040 Series: environmental life cycle assessment – principles and framework. Geneva: International Organization for Standardization; 2006.
- Laursen SE, Hanse, J, Knudsen HH, Wenzel H, Larsen HF, Kristensen FM. *EDIPTEX – Environmental Assessment of Textiles. Working Report No. 24*; 2000.
- Lehrburger C, Mullne J, Jones CV. *Diapers: Environmental Impacts and Lifecycle Analysis. Report to National Association of Diaper Services*; 1991.
- Little AD. *Disposable Versus Reusable Diapers: Health, Environmental and Economic Comparisons. Report to Proctor & Gamble*; 1990.
- Lo SC, Ma H, Lo SL. Quantifying and reducing uncertainty in life cycle assessment using the Bayesian Monte Carlo method. *Science of the Total Environment.* 2005;340:23–33.
- LRC-IES. *General Guide for Life cycle Assessment – Detailed Guidance. EUR 24708 EN – 2010.* 1st ed. European Commission Joint Research Centre and Institute for Environment and Sustainability; 2010.
- Nieminen E, Linkeb M, Tobler M, Vander Beked B. EU COST Action 628: life cycle assessment (LCA) of textile products, eco-efficiency and definition of best available technology (BAT) of textile processing. *Journal of Cleaner Production.* 2007;15(13–14):1259–1270.
- O'Brien KR, Olive R, Hsu YC, Bell R, Morris L, Kendall N. Life cycle assessment: reusable and disposable nappies in Australia. In: *6th Australian Life Cycle Assessment Society Conference 2009*, Melbourne (1–14), 17–19 February 2009.
- Pekala LM, Tan RR, Foo DC, Jezowski JM. Optimal energy planning models with carbon footprint constraints. *Applied Energy.* 2010;87:1–2.
- Sauer BJ, Hilderbrandt CC, Franklin WE, Hunt RG. Resource and environmental profile analysis of children's diaper systems. *Environmental Toxicology and Chemistry.* 1994; 13(6):1003–1009.
- Seuring S, Goldbach M. Managing sustainability performance in the textile chain. In: Wagner M, Schaltegger S, Wehrmeyer W, eds. *Sustainable Performance and Business Competitiveness.* Sheffield: Greenleaf; 2004.
- Sundarakani B, Souza RD, Goh M, Wagner SM, Manikandan S. Modeling carbon footprints across the supply chain. *International Journal of Production Economics.* 2010:1–4.
- Vizcarra AT, Lo KV, Liao PH. A life cycle inventory of baby diapers subject to Canadian conditions. *Environmental Toxicology and Chemistry.* 1994;13(10):1707–1716.
- Weidema BP, Wesnaes MS. Data quality management for life cycle inventories – an example of using data quality indicators. *Journal of Cleaner Production.* 1996;4(3–4): 167–174.

9.12 References

1. *The Carbon Footprint of a Cotton T-Shirt*. Executive Summary. UK: Continental Clothing Co. Ltd; March 2009. www.continentalclothing.com. Downloaded at: files. continentalclothing.com/press/LCA%20Executive%20Summary.doc.
2. Institute for Sustainable Resources, Queensland University of Technology. *Life Cycle Assessment of a 100% Australian-Cotton T-Shirt*; 2009. Available at: http://cottonaustralia. com.au/uploads/publications/Case_Study_Life_Cycle_Assessment. Peter_Grace.pdf.
3. Nagel, Schoeman, Yadavalli. Exploring ways to reduce the greenhouse gas emissions in the textile supply chain. In: *SAIIE Conference Systems, SAIIE 2010*; 2010. Available at: http:// www.saiie.co.za/ocs/index.php/saiie/SAIIE10/paper/view/161/81.
4. Nagel S. *Exploring Ways to Reduce Greenhouse Gas Emissions in the Textile Supply Chain* [Bachelors of Industrial Engineering thesis]. Faculty of Engineering, Built Environment and Information Technology, University of Pretoria; October 2010.
5. Steinberger JK, Friot D, Jolliet O, Erkman S. A spatially explicit life cycle inventory of the global textile chain. *International Journal of Life Cycle Assessment*. 2009;14:443–455.
6. *A Product Life Cycle Approach to Sustainability*. San Francisco, CA: Levi Strauss & Co.; March 2009. Available at: http://www.levistrauss.com/sites/default/files/librarydocument/ 2010/4/Product_Lifecyle_Assessment.pdf.
7. *An Environmental Product Declaration of Jeans*. Bio Intelligence Service, ADEME; 2006. Available at: www.ademe.fr/internet/eco-jean/EPD_en_jeans_v2.pdf.
8. Bevilacqua M, et al. Chapter 10 in case study: a carbon footprint analysis in textile supply chain. In: *Design for Environment as a Tool for the Development of a Sustainable Supply Chain*. Springer-Verlag London Limited; 2012:219–239. https://doi.org/10.1007/978-1-4471-2461-0_10.
9. Smith GG, Barker RH. Life cycle analysis of a polyester garment. *Resources, Conservation and Recycling*. 1995;14:233–249.
10. The Linen Shirt Eco-Profile, Press Release. *Masters of Linen; Eco-Profile of a Linen Shirt and Comparative Analysis of Linen and Cotton Shirts*. Bio Intelligence Service; February 2008. Available at: http://www.saneco.com/IMG/pdf/linen_shirt_eco-profile.pdf.
11. Ponder CS. *Cradle-to-Use Life Cycle Inventory of Medical Gowns, Sustainable Textiles and Medical Protections Conference*. PowerPoint Presentation Slides. UC Davis; June 1, 2010. Available at: http://nsf-muses.ucdavis.edu/pdf/STAMPconference/CPonder%20-Gown%20Comparison%20v2.pdf.
12. Ponder CS, Overcash M. Life cycle assessment of healthcare garments. In: *LCA/LCM 2007 Conference, Portland, Oregon*; October 4, 2007. PowerPoint Presentation Slides. Available at: http://lcacenter.org/inlca2007/presentations/75.pdf.
13. Mikusinska M. *Comparative Life Cycle Assessment of Surgical Scrub Suits: The Case of Reusable and Disposable Scrubs Used in Swedish Healthcare* [Degree Project Thesis, Second Level, SoM EX 2012-48]. Stockholm: Department of Urban Planning and Environment, Division of Environmental Strategies Research – FMS, KTH, Royal Institute of Technology; 2012. Available at: http://kth.diva-portal.org/smash/get/diva2:605294/ FULLTEXT01.pdf.
14. Sau-Fun Ng F, Muthu SS, Li Y, Chi-Leung Hui P. A critical review on life cycle assessment studies of diapers. *Critical Reviews in Environmental Science and Technology*. 2013; 43(16):1795–1822.
15. Aumônier S, Collins M. *Life Cycle Assessment of Disposable and Reusable Nappies in the UK*. UK: Environment Agency; 2005.

16. Aumônier S, Collins M, Garrett P. *An Updated Lifecycle Assessment Study for Disposable and Reusable Nappies*. Science report SC010018/SR2. Almondsbury, Bristol, UK: Environment Agency; 2008.
17. Muthu SS, Ng FSF, Li Y, Hui PCL, Guo Y. Carbon and eco-footprints of adult incontinence products. *Fibers and Polymers*. 2013;14(10):1776−1781.
18. Velden NM, Patel MK, Vogtländer JG. LCA benchmarking study on textiles made of cotton, polyester, nylon, acryl, or elastane. *International Journal of Life Cycle Assessment*. 2014;19:331−356.
19. Vogtländer JG, Bijma A. The virtual pollution prevention costs '99: a single LCA-based indicator for emissions. *International Journal of Life Cycle Assessment*. 2000;5(2): 113−124.
20. Vogtländer JG, Brezet HC, Hendriks CF. The virtual ecocosts '99: a single LCA-based indicator for sustainability and the ecocosts/value ratio (EVR) model for economic allocation. *International Journal of Life Cycle Assessment*. 2001;6(3):157−166.
21. Zhang Y, Liu X, Xiao R, et al. *International Journal of Life Cycle Assessment*. 2015;20: 994. https://doi.org/10.1007/s11367-015-0889-4.

Assessing the environmental impact of textiles: summary and conclusions

10

10.1 Introduction

It is well known that textile products are responsible for significant environmental impacts across the globe. The effective measurement and mitigation of these impacts are vital for the preservation of our planet. This book has discussed a range of topics relating to the assessment of the environmental impact of textiles.

The first chapter reviewed the entire textiles and clothing supply chain in terms of the environmental issues related to different fibre types, yarn, fabric and apparel production methods. The chapter introduced the concept of a product's life cycle and discussed the potential environmental impacts of each phase of the life cycle.

The second chapter discussed the various methods available today to quantify the environmental impacts of textile products and processes. Throughout the book, this was discussed in terms of the life cycle assessment (LCA) approach. LCA can be used to measure various environmental indicators, and this chapter threw light on the key differences between them. It also discussed the environmental legislation applicable both generally and in various parts of the world, such as the United States and the European Union.

Chapter 3 focused on the concept of the product carbon footprint (PCF) and the measurement techniques and standards available for the quantification of greenhouse gas emissions. The key challenges in modelling the PCF were highlighted, especially the problems of including the entire supply chain.

The next two chapters discussed the concepts, methods, standards and modelling techniques for assessing the energy and water footprint (Chapter 4) and ecological footprint (Chapter 5) of textile products. These chapters also discussed the applications of these concepts to the key stages of the textile and clothing supply chain and highlighted the difficulties in modelling them in a detailed way.

Chapter 6 discussed the history, concept, methods and standards of LCA, presenting a detailed and comprehensive picture of LCA methodology, its advantages and disadvantages and its four phases, as defined by the ISO 1404X series of standards. Each step was discussed in detail, and the basics of LCA modelling of, particularly, textile products were explained.

Chapter 7, one of the most important chapters in this book, dealt with the practical aspects of modelling LCA and PCF for textile products. It explained the ways of creating the models for textiles and clothing products to measure life cycle impacts/carbon/ecological footprint or any other indicator, for that matter. This chapter also highlighted the limitations and strengths of the publicly available databases and datasets for textiles. The various challenges and difficulties involved in modelling LCA and PCF for textile products were discussed in detail.

Assessing the Environmental Impact of Textiles and the Clothing Supply Chain
https://doi.org/10.1016/B978-0-12-819783-7.00010-7

Chapter 8 explained the end-of-life management of textile products highlighting the various end-of-life destinations and their implications for the environment. The concept of biodegradation and its applicability to textile products was also discussed.

Chapter 9 dealt with the measurement of environmental impacts of textiles and clothing products in practice through case studies. Some of the most familiar and important categories of textiles and clothing products were chosen, and LCA and PCF measurement of those products was discussed in detail to help the reader gain a better understanding of the LCA and PCF of textile products. The case studies included cotton T-shirts, denim jeans, a polyester blouse, a polyester jacket, a woollen sweater, a linen shirt, medical textile products and nonwoven nappies (diapers). Chapter 9 also dealt with the consumer use phase as part of the LCA and PCF of textiles and clothing.

Thus a complete journey through the environmental assessment of the supply chain of textiles and clothing, with special reference to LCA and PCF, has been presented in the previous nine chapters. This chapter will now briefly discuss the choice of which assessment methods to use and when. It will also highlight current problems and future trends in the assessment of LCA and PCF for textile products.

10.2 Which assessment methods to use and when

Different methods are available today to measure the environmental impacts of textiles and clothing products, including carbon footprint, ecological footprint, water footprint, energy footprint and multiple indicators using LCA. Even if single indicators such as carbon or ecological footprints are to be used, still it is advisable and better to follow the LCA approach to quantify any particular indicator. Any indicator can be studied once the basic LCA model is established, as shown earlier. Depending on the product profile and the nature and scope of the assessment required, it is up to the LCA expert to choose the most appropriate indicator (or indicators). For instance, for a cotton fabric LCA involving the farming system, it is not advisable to neglect the study of eutrophication or water consumption.

As discussed in Chapter 6, various impact assessment methods are available with any LCA software these days, and each method has its unique indicators. At times, it is also essential to choose various indicators from different methods to present a comprehensive view of the environmental impacts associated with the product system in question. It is also desirable to present multiple indicators for all the life cycle phases of a product, if its eco-profiles are to reliably present the overall picture. It is also important to extend the boundary of the assessment from cradle to grave for any indicator; this is always more desirable than cradle to gate. Most current standards stress the importance of a full LCA rather than the partial LCA of products.

10.3 Current problems in environmental assessment

As mentioned in a previous chapter, the concepts of LCA and PCF are still in their infancy as far as textile products are concerned, although they are well established in other industrial sectors. In spite of being a basic consumer product, it is both surprising and unfortunate that there have not been more studies on textiles or clothing products. There are at present only a few studies in a small number of product categories, and this is a major omission that needs to be addressed on an urgent basis. Compared to the timeline when the first edition of this book was published, there are some more products assessed for LCA and PCF, in which some of them are mentioned in the second edition of this book. Still I can see a lot of LCA studies can be developed for textile sector and a holistic approach of assessment is the need of the hour for various textile processes, fibres, materials and products. It is also very important to note that many LCA studies are emerging and being published (publicly available in the internet, social media) without having a critical review. Only a few studies, except being published in international peer-reviewed journals, include a critical review part. Even within the available studies, lack of consistency is a major problem, making it very difficult to compare results and establish the trends.[1] There is no consistency in terms of selection of indicators, assumptions and reporting of results; even for the same product, different studies chose differing functional units. No common methodology is followed in these studies, and in some of the studies the functional unit is not even defined properly. Most of the studies published to date do not explain their underlying methodology, without which it is difficult to make proper use of the data or to replicate them. All these issues can be seen in the case studies discussed in Chapter 9. Many PCF studies do not explain the emission factors chosen and their sources, and this is true for LCA studies, where life cycle inventory is not explained, with appropriate sources. LCA studies are still not available for many important fibres, fibre blends and the most common textile product categories.

Eco-profiling is inevitable today for any industrial sector, and an eco-profile is publicly available for most products from various industrial sectors. However, there are very few eco-profiles available for most of the fibres, materials and products in the textile sector. The same is true for Product Category Rules (PCRs) for textiles and clothing products. Owing to these factors, together with the scarcity of datasets for textile materials, a textile researcher finds it difficult to find information for any studies. This scenario has to be changed as quickly as possible.

As shown in earlier chapters, there are still a lot of grey areas in the environmental assessment of textiles. Many indicators have not been explored for textile products, even though they are very much applicable to the textiles and clothing sector. The same is the case for some of the life cycle phases of textile products. There is a scarcity of information on the recycling of different fibre group materials and their environmental implications,[2] as well as on the biodegradation of different textile materials.

As discussed in detail in Chapters 3–7, for the assessment of any indicator including LCA, there are a lot of challenges and difficulties to be faced, and some of the most important issues are

- defining a proper functional unit'
- setting of stages and boundaries of measurement;
- building proper justifiable assumptions;
- technical know-how in terms of creating LCA models;

- verifiability of the modelled results;
- choice of a suitable methodology/standard for the measurement of any indicator;
- selecting appropriate emission factors for PCF studies and confirming their suitability;
- properly selecting data and data quality and data uncertainty;
- technical know-how in data collection and the filling of gaps with appropriate data;
- availability of suitable datasets from databases and literatures;
- selection and availability of appropriate secondary data, if primary data are not available;
- conversion of data from secondary sources to suit the current study;
- missing critical review of the LCA study;
- tackling methodological complexities, such as allocation of products and co-products, and dealing with cut-off rules for the study.

10.4 Future trends

Future trends must address the issues pointed out many times in earlier chapters in this book, and also highlighted in this chapter. It is pivotal to have a common methodology for various textile products (groupwise), along with LCA-based information for various life cycle phases. The ways of reducing the environmental impacts arising from textiles and clothing must be addressed, and research on textile sustainability should be increased to address all these issues. Some of the main areas to be focused on are as follows:

- Although it is difficult to compare different products assessed by different studies, due to factors such as geographical location and energy carriers used, a common methodology must be developed for textiles and clothing. LCA or PCF studies that arise from a particular area/location can at least be compared or referenced.
- Functional units, assumptions and other methodological issues should be fixed for textiles and clothing products, and all studies should follow the same approach. Development of PCRs for all textile product categories should be able to solve this issue. Different life cycle phases need to be discussed for various textile fibres, materials and clothing products.
- A dedicated database should be created for textiles and clothing products.
- A product library addressing the life cycle impact issues for various life cycle phases of textile products must be created and made publicly available.
- A common tool or platform must be developed to indicate the average lifetime of products in relation to the functional limit of different fibre materials, including the details of how to deal with 'use' phase scenarios for different clothing items.
- A unified methodological base must be created for a range of textile products (and hence for different product categories), and all textile studies must follow it.
- Environmental impact information on different textile fibres, intermediate forms of textiles such as yarns, and fabrics made by various methods/techniques must be made available. Such information must be scientific and must come from scientific models, and not from anecdotes and unsupported 'facts'.
- Detailed information on various environmental aspects such as energy use, water consumption, carbon footprint, ecological footprint and so on should be available for various fibres and different textile products, encompassing the complete life cycle phases of textile products. This can be geographical location specific.

- Detailed information on how to treat different garments must be displayed, probably in the form of a tag. The scientifically established best ways of caring and maintaining a typical textile product must be suggested to the consumers. As the consumer use phase was often found to be the most detrimental one, this has to be done to minimize the life cycle impacts that arise from clothing products.
- As a result of many LCA studies, information on various environmental impacts must be displayed on the products, along with the care labels.
- LCA studies must be made on advanced textile applications such as nano textile materials and plasma-treated materials.
- Potential environmental impacts applicable to the textile sector must be determined for different textile products, and detailed notes on their assessment methods and possible mitigation measures must be reported.
- Suitable environmental impact indicators must be developed for textile and clothing products, along with a suitable methodology dedicated to this sector. This methodology should have the flexibility to normalize the impacts for various countries around the world.
- Sufficient information must be provided in terms of the scientifically established benefits to be gained from reusing/recycling different clothing products. Also, the minimum number of times a clothing product should be used/reused – to benefit environmental preservation and achieve sustainable development – must be declared.
- Product-longevity-included studies based on actual scenarios rather than assumptions on functional life to define a functional unit are still lacking in the scientific fraternity and many such studies are required.
- A social LCA methodology has been developed, separate from LCA. Social LCA has been conducted in some industrial sectors and is lately gaining importance. Social aspects are quite common in the textiles sector, but so far, this has not been addressed by any study. Suitable social impact indicators applicable to the clothing sector must be developed, along with the methodology to assess them.
- Another uncharted area is life cycle costing in the textiles and clothing sector. The eco-design approach should be followed for many textile and clothing products, and details pertaining to eco-design and re-design should be made available for each product category.
- Concepts such as eco-functional assessment methodologies[3,4] must be utilized to evaluate various textile products, and eco-functional indices must be made available for different textile and clothing products.
- Cradle-to-cradle kind of approaches[5] must be tried for textile products with a systematic scientific approach, which is the need of the hour. Although they are seen to be coming up, many holistic approach studies such as LCA need to be done for many textile products.
- Details of open- and closed-loop recycling techniques, with the corresponding life cycle impacts/benefits, must be investigated for various clothing products, and the results should become the basis of suitable recommendations.
- Process-based LCA studies must be completed for different processes, materials and technologies involved in the textile supply chain, along with recommendations to mitigate their impact.

10.5 Sources of further information and advice

- Dahllöf L. Life Cycle Assessment (LCA) applied in the textile sector: the usefulness, limitations and methodological issues – a literature review. In: *Environmental Systems Analysis*. Göteborg, Sweden: Chalmers University of Technology; 2003. ESA-R2003:9, ISSN:1404-8167.

- Dahllöf L. Methodological issues in the LCA procedure for the textile sector: a case study concerning fabric for a sofa. In: *Environmental Systems Analysis*. Chalmers University of Technology; 2004. ESA-Report 2004:7, ISSN:1404-8167.
- McDonough W, Braungart M. *The Upcycle: Beyond Sustainability—Designing for Abundance*. North Point Press; 2013.
- Muthu SS. *Eco-functional Assessment of Grocery Shopping Bags* [Doctor of Philosophy Thesis]. Hong Kong: Institute of Textiles and Clothing, The Hong Kong Polytechnic University; 2012.
- Muthu SS. *Handbook of Life Cycle Assessment (LCA) of Textiles and Clothing*. Woodhead Publishing; 2015.

10.6 References

1. Chapman A. *Review of Life Cycle Assessments of Clothing*. A Report for Mistra Future Fashion. Oakdene Hollins Research & Consulting; July 2010. File Reference Number: MIST01 232 LCA review.doc. Available at: http://www.oakdenehollins.co.uk/media/232/2010_mistra_review_of_life_cycle_assessments_of_clothing.pdf.
2. Muthu SS, Li Y, Hu JY, Li Z. Carbon footprint reduction in the textile process chain: recycling of textile materials. *Fibers and Polymers*. 2012;13(8):1065–1070.
3. Muthu SS, Li Y, Hu J-Y, Mok P-Y, Lin M. Modelling and quantification of eco-functional Index: the concept and applications of eco-functional assessment. *Ecological Indicators*. March 2013;26:33–43.
4. Muthu SS, Li Y, Hu JY, Mok PY. Eco-functional assessment combined with life cycle analysis: concept and applications. *Energy, Education, Science and Technology*. 2012;29(1): 435–450.
5. McDonough W, Braungart M. *Cradle to Cradle: Remaking the Way We Make Things*. North Point Press; 2002.

Index

'*Note*: Page numbers followed by "f" indicate figures and "t" indicates tables.'

J

K

L

FSC
www.fsc.org
MIX
Papier | Fördert gute Waldnutzung
FSC® C083411

Kolibri 360 GmbH
Ferdinand-Jühlke-Straße 7
99095 Erfurt, Deutschland
produktsicherheit@kolibri360.de